Biology Is Technology

BIOLOGY IS TECHNOLOGY

The Promise, Peril, and New Business

of Engineering Life

ROBERT H. CARLSON

HARVARD UNIVERSITY PRESS

Cambridge, Massachusetts

London, England

2010

For Pascale

Library of Congress Cataloging-in-Publication Data

Carlson, Robert.
  Biology is technology : the promise, peril, and new business of engineering life / Robert
Carlson.
    p. cm.
  Includes bibliographical references and index.
  ISBN 978-0-674-03544-7 (alk. paper)
  1. Biotechnology.   2. Biology—Philosophy.   3. Bioethics.   I. Title.
  TP248.2.C37 2010
  660.6—dc22   2009029637

Contents

Acknowledgments

THIS BOOK is the sum of more than a decade of discussions, research, and travel. The effort was made possible only through the generosity and patience of friends, colleagues, and family.

The first draft was written while I was a Visiting Scholar in the Comparative History of Ideas Program at the University of Washington; many thanks to John Toews and the late Jim Clowes.

I am especially grateful to Dianne Carlson, Eric Carlson, Drew Endy, Sarah Keller, Rik Wehbring, and anonymous reviewers for reading drafts, providing invaluable comments, and for asking hard questions.

My journey into thinking about biology's role in the economy benefited at every step from conversations with my colleagues Stephen Aldrich and James Newcomb at Bio Economic Research Associates (Bio-era). Our intellectual travels through biofuels, pandemics, vaccines, commodities trading, the oil and gas business, and international finance, supplemented by our real-world travels through North America, Asia, and Europe, provided an unparalleled education.

I owe thanks to many others for conversations, however brief or involved, that provided welcome critique and insight, including Drew Endy, Stewart Brand, Kevin Kelly, Freeman Dyson, Emre Aksay, W. Brian Arthur, Ralph Baric, Sydney Brenner, Roger Brent, Ian Burbulis, Charles Cantor, Denise Caruso, Jamais Cascio, Joseph Chao, Napier Collyns, David Grewal, Lauren Ha, Andrew Hessel, Janet Hope, Richard Jefferson, Tom Kalil, John Koschwanez, Tom Knight, Ed Lazowska, Emily Levine, Barry Lutz, John Mulligan, Oliver Morton, Kenneth Oye, Bernardo Peixoto, Arti Rai, Brad Smith, Richard Yu, Steve Weber, and last, but certainly not least, Ben and Margit Rankin. I thank my editor, Michael Fisher, for his patience over many years.

CHAPTER ONE

What Is Biology?

Biology is technology. Biology is the *oldest* technology. Throughout the history of life on Earth, organisms have made use of each other in sophisticated ways. Early on in this history, the ancestors of both plants and animals co-opted free-living organisms that became the subcellular components now called chloroplasts and mitochondria. These bits of technology provide energy to their host cells and thereby underpin the majority of life on this planet.

It's a familiar story: plants, algae, and cyanobacteria use sunlight to convert carbon dioxide into oxygen. Those organisms also serve as food for a vast pyramid of herbivores and carnivores, all of whom produce carbon dioxide and other wastes that plants then use as resources.

Interactions between organisms constitute a global natural economy that moves resources at scales from the molecular to the macroscopic, from a few nanometers (10^{-9}) to many megameters (10^6). Humans have always explicitly relied on this biological economy to provide food, oxygen, and other services. Until recently, our industrial economy relied primarily on nonbiological technologies; the industrial revolution was built primarily on fire, minerals, and chemistry. Now, however, our economy appears to be changing rapidly, incorporating and relying upon new organisms whose genomes have been modified through the application of human effort and ingenuity.

In 2007, revenues in the United States resulting from genetic modification of biological systems were the equivalent of almost 2 percent of gross domestic product (GDP). The total includes all the products we include under the moniker "biotechnology"—drugs, crops, materials, industrial enzymes, and fuels (see Chapter 11). Compare that 2 percent to the percent added in 2007 to GDP from the following sectors: mining, 2 percent; construction, 4.1 percent; information and broadcasting, 4.7 percent; all of manufacturing, 11.7 percent; transportation and warehousing, 2.9 percent; finance, 20.7 percent; and all of government, 12.6 percent.[1] (One might expect the contribution from finance to be somewhat smaller in the future.)

While still modest in size compared with other sectors, biotech revenues are growing as fast as 20 percent annually. Moreover, the sector is extremely productive. During the years 2000–2007, the U.S. economy expanded by about $4 trillion, and biotech revenues grew by almost $200 billion. Biotechnology companies currently employ about 250,000 people in the United States, out of a total labor force of 150 million.[2] Therefore, less than one-sixth of 1 percent of the national workforce generated approximately 5 percent of U.S. GDP growth during those seven years. Despite the fact that the underlying technology is presently immature compared with other sectors of the economy, current biotechnology demonstrates impressive and disproportionate economic performance.

Rapid revenue growth in the sector is the result of new products that create new markets, such as drugs and enzymes that help produce fuels. It also comes from displacing products made using older industrial methods. Bioplastics that started entering the market in 2007 and 2008 appear to require substantially less energy to produce than their petroleum-based equivalents.

Yet, as with any other technology, biological technologies are subject to the hard realities of the market. New products may fail for many reasons, including both overoptimistic assessment of technical capabilities and customer inertia. In addition, biotechnology must compete with alternate methods of producing materials or fuels, methods that may have a century's head start. Biological production must also compete for feedstocks with other human uses of those feedstocks, as is now occurring in the commercialization of first-generation biofuels produced from sugar, corn, and vegetable oil. It is no surprise that many biofuel producers are presently caught up in the collision between food and fuel; crops, and the resources used to grow them, are likely to have higher value as food than as fuel.

The economic system that governs these products is today primarily composed of interconnected marketplaces, full of businesses large and small. Those markets are increasingly global, and the flow of information is at least as important as the flow of physical goods. Technology supports the spread of those markets, and technology is the subject of many of those markets. New technologies provide opportunities to expand markets or launch entirely new ones. Here I use the word "market" in the broadest sense, which Wikipedia (presently) defines as "any of a variety of different systems, institutions, procedures, social relations and infrastructures whereby persons are coordinated, goods and services are exchanged and which form part of the economy."[3] I do not mean any particular market, nor necessarily the "free market," nor any particular set of transactions governed by any particular set of rules and denominated in any particular currency.

In general, as we shall see, there is no reason to think that any country's lead in developing or using biological technologies will be maintained for long. Nor will the culture and experience of any given country dominate the discussion. Access to biological technologies is already ubiquitous around the globe. Many countries are investing heavily to build domestic capabilities with the specific aim of improving health care, providing fuels and materials, and increasing crop yields.

Research efforts are now accelerating, aided by rapid advances in the technology we use to manipulate biological systems. It is already possible to convert genetic information into electronic form and back again with unprecedented ease. This capability provides for an element of digital design in biological engineering that has not heretofore been available. More important, as measured by changes in commercial cost and productivity, the technology we use to manipulate biological systems is now experiencing the same rapid improvement that has produced today's computers, cars, and airplanes. This is evidence that real change is occurring in the technologies underlying the coming bioeconomy.

The influence of exponentially improving biological technologies is only just now starting to be felt. Today writing a gene from scratch within a few weeks costs a few thousand dollars. In five to ten years that amount should pay for much larger constructs, perhaps a brand-new viral or microbial genome. Gene and genome-synthesis projects of this larger scale have already been demonstrated as academic projects. When such activity becomes commercially viable, a synthetic genome could be used to build an organism that produces fuel, or a new plastic, or a vaccine to combat the outbreak of a new infectious disease.

This book is an attempt to describe a change in technology that has demonstrably profound social and economic implications. Some parts of the story that follows I know very well, either because I was fortunate to witness events or because I was in a position to participate. Other parts of the story come in because I had to learn something new while attempting to paint a picture of the future. Delving into details is necessary in places in order to appreciate the complexity of biological systems, the challenge of engineering those systems, and the implications of that technology for public policy, safety, and security. Whatever else the reader takes from this book, the most important lesson is that the story is incomplete. Biology is technology, and as with any other technology, it is not possible to predict exactly where the project will go. But we can at least start with where that technology has been.

Engineering Organisms Is Difficult, for Now

Explicit "hands-on" molecular manipulation of genomes began only in the mid-1970s, and we are still learning the ropes. Most genetically modified systems do not yet work entirely as planned. Biological engineering as practiced today proceeds by fits and starts, and most products on the market now result from a process that remains dominated by trial and error. The primary reason that the engineering of new organisms has been slow in coming is that simply understanding naturally occurring organisms remains hard.

The initial phase of biological engineering, covering the last thirty years or so, coincided with efforts to describe the fundamentals of molecular biology. In that time we moved from discovering the number of genes in the human genome to building automated machines that read entire microbial genomes during a lunch break. Science has accumulated enough knowledge to support basic genetic changes to microbes and plants; those changes enable a wide range of first-generation products.

What was cutting-edge technology three decades ago is today routine in university lab courses and has already been included in the curricula of many high schools. While simple modifications of single-celled organisms are now commonplace, the scientific frontier has, of course, moved. Today, academic and industrial researchers alike are working with multicellular organisms and contending with the attendant increase in biochemical and developmental complexity.

And yet progress can appear slow, particularly to those who have followed the information technology revolution. Governments and big business once dominated computing. Today, entrepreneurs and garage hackers play leading roles in developing computing technologies and products, both hardware and software.

Thus, Stewart Brand, founder of the *Whole Earth Catalog*, organizer of the first Hackers' Conference in 1984, and cofounder of the Whole Earth

'Lectronic Link (WELL) and the Global Business Network, wonders: "Where are all the green biotech hackers?"[4] To which I answer: "They are coming." The tools necessary to understand existing systems and build new ones are improving rapidly. As I will discuss in Chapter 6, the costs of reading and writing new genes and genomes are falling by a factor of two every eighteen to twenty-four months, and productivity in reading and writing is independently doubling at a similar rate. We are just now emerging from the "slow" part of the curves, by which I mean that the cost and productivity of these technologies are now enabling enormous discovery and innovation. Consequently, access to technology is also accelerating. "Garage biology" is here already; in Chapter 12 I share a bit of my own experience sorting out how much innovation is possible in this context.

Public Expectations for Advances in Biological Technologies

The new knowledge and inventions that science provides can take many decades to become tools and products—things people can buy and use—that generate value or influence the human condition. That influence is, of course, not uniformly beneficial. But we generally cannot know whether a technology is, on balance, either valuable or beneficial until it is tested by actual use. It is in this context that we must examine new biological technologies.

Biological technologies are subject to both unreasonable expectations and irrational fear. Practitioners and policy makers alike must contend with demands by the citizenry to maximize benefit speedily while minimizing risk absolutely. These demands cannot be met simultaneously and, in many cases, may be mutually exclusive. This tension then produces an environment that threatens much-needed innovation, as I argue in detail in the latter half of this book.

Often sheer surprise can play as great a role in public responses as the science itself. By the nature of the scientific process, most results reported in the press are already behind the state of the art. That is, while ideally science is news of the future, the press actually reports the past. Scientific papers are submitted for publication months after work is finished, go through a review and editing process consuming additional months, and finally appear in print many months after that, all the while being surpassed by ongoing research. Given the pace of technological improvement and consequent increased capabilities in the laboratory, more and more new science is being squeezed into the time between discovery and publication of old results.

Only if we recognize that organisms and their constituent parts are engineerable components of larger systems will we grasp the promise and the peril of biological technologies. Conversely, failing in this recognition will cloud our ability to properly assess the opportunity, and the threat, posed by rapid changes in our ability to modify biological systems.

We are in the midst of realizing capabilities first forecast more than fifty years ago. The development of X-ray crystallography and nuclear magnetic resonance in the decades before 1950 opened a window to the molecular world, providing a direct look at the structure of natural and synthetic materials. During that same time period the elucidation of information theory, cybernetics, and basic computational principles set the stage for today's manipulation of information. Biology is the fusion of these two worlds, in which the composition and structure of matter determines its information content and computational capabilities. This description may also be applied to computers, but biology is in addition a state of matter, if you will, that is capable of self-editing and self-propagation. It is no surprise, then, that given our improving abilities to measure and manipulate molecules, on the one hand, and to apply powerful computational techniques to understand their behavior, on the other, biology is today consuming considerable attention. Today's science and technology provide a mere glimpse of what is in store, and we should think carefully about what may happen just down the road.

What Is Biological Engineering?

As we shall see in the following chapters, the development of new mathematical, computational, and laboratory tools will facilitate the building of things with biological pieces—indeed, the engineering of new biological artifacts—up to and including new organisms and ecosystems. The rest of this book explores how this may transpire. But first we have to understand what engineering is.

Aeronautical engineering, in particular, serves as an excellent metaphor when considering the project of building novel biological systems. Successful aeronautical engineers do not attempt to build aircraft with the complexity of a hawk, a hummingbird, or even a moth; they succeed by first reducing complexity to eliminate all the mechanisms they are unable to understand and reproduce. In comparison, even the simplest cell contains far more knobs, bells, and whistles than we can presently understand. No biological engineer will succeed in building a system de novo until most of that complexity is stripped away, leaving only the barest essentials.

A fundamental transformation occurred in heavier-than-air-flight, starting about 1880. The history of early aviation was full of fantastical machines that might as well live in myth, because they never flew. The vast majority of those failed aircraft could, in fact, never leave the ground. They were more the product of imagination and optimism than of concrete knowledge of the physics of flight or, more important, practical *experience* with flight.

In about 1880, Louis-Pierre Mouillard, a Frenchman living in Cairo, suggested that rather than merely slapping an engine on a pair of wings and hoping to be pulled into the air, humans would achieve powered flight only through study of the practical principles of flight. And that is just the way it worked. Aviation pioneers made it into the air through careful observation and through practice. These achievements were followed by decades of refinement of empirical design rules, which were only slowly displaced in the design process by quantitative and predictive models of wings and engines and control systems. Modern aircraft are the result of this process of learning to fly.

And so it goes with virtually every other human technology, from cars to computers to buildings to ships, dams, and bridges. Before inanimate objects are constructed in the world today, they are almost uniformly first constructed virtually—built and tested using sophisticated mathematical models. Analogous models are now being developed for simple biological systems, and this effort requires molecules that behave in understandable and predictable ways. The best way to understand how biological technology is changing is by starting with another metaphor, one that relies on the best toy ever devised: LEGOS.

CHAPTER TWO

Building with Biological Parts

B UILDING WITH LEGOS is an excellent metaphor for future building with biology. The utility and unifying feature of LEGOS, Tinkertoys, Erector Sets, Zoob, or Tente is that the pieces fit together in very understandable and defined ways. This is not to say they are inflexible—with a little imagination extraordinary structures can be built from LEGOS and the other systems of parts. But it is easy to see how two bricks (or any of the other newfangled shapes) will fit together just by looking at them.

The primary reason that we fret about bioterror and bioerror is that in human hands biological components aren't yet LEGOS. But they will be, someday relatively soon.

Every Piece Has Its Purpose

LEGOS are so broadly used and understood in our culture that they are employed to advertise other products. There is a metaphorically and visually elegant television commercial for the Honda Element that opens with an image of a single LEGO-style brick. That computer-generated brick is soon joined by a myriad of others, with different shapes and colors, each flying into its assigned place with inspiring precision. The wheels and floor of the vehicle quickly take shape, followed by clever folding seats, and finally the sides and roof. It is clear that all the parts of this building-block car are carefully conceived, designed, and built. At the end of the spot, a smooth voice-over brings it all together: "Every piece has its purpose."[1]

The implication is clear: Honda has built a new vehicle, and each and every part does exactly what it is supposed to. We can feel confident that each part performs precisely as intended. This, after all, is a modern car and the fruit of sophisticated modern engineering practice.

There are many unspoken assumptions built into this representation of engineering. Among the most important are that (1) all the parts are the result of a careful design process, (2) the parts can be constructed to function according to the design, and (3) when assembled from those parts, the resulting whole actually behaves as predicted by the design.

The experience of everyone who watches the commercial contributes to the communication of these unspoken assumptions. Not only do we the viewers have considerable exposure to other products of this engineering process, but, given the number of Honda cars and trucks on the road, many of us demonstrate confidence in the engineering and manufacturing prowess of Honda in particular.

Just as the broad public understanding of LEGOS can be used to imbue a sense of careful design and manufacturing into a new Honda, the notion that the products of modern engineering are safe and predictable can be used to sell other technologies.

Unfortunately, in comparison, current genetic "engineering" techniques are quite primitive, akin to swapping random parts between cars to produce a better car. Biological engineering in general does not yet exist in the same way that electrical, mechanical, and aeronautical engineering do. Mature engineering fields rely on computer-aided design tools—software packages like SolidWorks for mechanical engineering and Spice or Verilog for circuit simulation—that are based upon predictive models. These predictive models are constructed using a quantitative understanding of how parts of cars and airplanes behave when assembled in the real world. Unlike the vast majority of modified biological systems, for which there are no design tools, the behavior of a finished engine or integrated circuit can be predicted from the behavior of a model, which today is universally determined using computer simulation.

Implementation of computer-based automotive, electronics, and aircraft design efforts is aided by standardized test and measurement gear, such as oscilloscopes, network analyzers, stress meters, pressure gauges, etc. The combination of these items with predictive models constitutes an engineering toolbox that enables the construction of physical objects. Facilitating this construction process, programs like SolidWorks and Spice can send instructions to automated manufacturing tools, turning design into artifact in relatively short order. Although biological raw materials are quite different from gears, engines, and circuits, biology will soon have its own engineering toolbox.

The development of relevant tools is already well under way. Technologies used to measure and manipulate molecules and cells will be critical components of the toolbox. Laboratory instruments such as DNA sequencers and synthesizers, which read and write genetic instructions, respectively, are plummeting in price while becoming exponentially more powerful. This technology is changing so rapidly that, within just a few years, the power of today's elite academic and industrial laboratories will be affordable and available to individuals.

Toward Biological Composability

There are different ways of using biology as technology. Farming and breeding are obviously fundamental technologies. Bioremediation is ever more widely used to clean up messes—made by other human technologies—through the clever selection of plants and animals that thrive on materials we now generally consider waste. Bacteria have been used to produce electricity from sewage, and, as I will explore in detail in Chapter 11, microorganisms are being genetically manipulated to produce fuels.[2] The products of genetically modified systems are already used in homes and businesses worldwide, from laundry enzymes to powerful new medicines. Recombinant DNA is the technology that makes this possible.

The phrase "recombinant DNA" is derived from the earliest techniques developed to directly manipulate the genomes of bacteria. Without delving into the details, suffice it to say for the moment that recombination is an ancient biological process, wherein two pieces of DNA with ends that have similar chemical structure are pasted together by the preexisting molecular machinery of the cell. After this process takes place, the "seam" between the two ends is effectively invisible. The resulting DNA is chemically indistinguishable from the original, save that the new sequence can now code for new instructions. This process—this *technology*—appears to have evolved early in the history of life on this planet, as a mechanism to repair broken DNA strands.

The technological aspects of biology thus go down to the very molecules that underlie life itself. Nucleic acids, assembled into chains, constitute the molecular repository of instructions required to build any organism. Deoxyribonucleic acid (DNA) is the primary storage medium for all organisms composed of cells, whereas viruses may use either DNA or ribonucleic acid

(RNA). While the DNA describing how to build "higher" organisms such as humans is largely static over the lifetime of an individual, bacteria may alter their DNA content through the exchange of small bits of DNA, arranged in stable, circular elements called plasmids. Plasmids are extraordinarily useful bits of technology, a kind of standard packaging for DNA, which I will return to time and again in this book. Plasmids can serve as the medium for the transfer between individuals of useful genes, such as those for antibiotic resistance, and can also be passed along to offspring. Humans have for decades made use of this packaging technology by applying a high-voltage jolt to open temporary pores in microbes through which plasmids can migrate. But it has only recently been demonstrated that humans merely reinvented this trick, known as electroporation. Certain soil-dwelling microbes are particularly likely to take up plasmids from the environment after lightning strikes, a trait that enhances the ability of the microbes to sample the diversity of DNA lying around in the environment and thus to pick up genes that, if they are fortunate, are useful.[3]

Yet more sophisticated are natural schemes to respond to specific threats in the environment. The genome of *Vibrio cholerae,* the organism that causes cholera, possesses in its chromosome another sort of DNA-packaging technology, integrated conjugative elements (ICEs), which contain genes that confer resistance to particular antibiotics. The remarkable thing about this technology is that exchange of these ICEs between bacteria is inhibited *except* when those antibiotics threaten the bacteria.[4]

There is a significant evolutionary advantage for microbes that develop mechanisms that allow them to defend against human attacks. Prior to 1993, the ICE in Asian *V. cholerae* strains that confers antibiotic resistance against ciprofloxacin (Cipro) was not found in nature. It is now present in almost all samples isolated from cholera victims in Asia. Of specific interest (and perhaps worry) in this case, the bacteria have developed a mechanism in which the presence of Cipro actually *promotes* the spread of genes that code for antibiotic resistance, which accounts for the wide distribution of Cipro-resistant cholera infections. This exchange of genetic material, while by no means intentionally managed by microbes, provides a set of tools that allow organisms to adapt to, and even manipulate, environments that would otherwise be fatal.

Ultimately, however, we humans are the most successful organisms on the planet when it comes to using biology as both tool and raw material. We have, intentionally or otherwise, manipulated many species of plants, animals, fungi, and bacteria for thousands of years. It is now clear that humans cultivated corn at least nine thousand years ago, selecting plants with useful (or tasty) random genetic changes, combining those mutations into a single plant via breeding, and then propagating its seeds.[5] Even with this relatively sophisticated genetic manipulation, we are certainly late to the game; other organisms have been at it far longer. Yet humans make use of a greater diversity of species than any other organism on Earth. This reliance has long nourished both the human body physical and the human body social. It is unlikely we will find an alternative to this habit anytime soon, even if we actively sought it.

Adaptation of all the "found technology" of molecular biology to human ends was first demonstrated in the early 1970s and was rapidly adopted to produce proteins on a commercial scale. The process works as follows: First, instructions for making a protein are inserted into a cell via recombination. Those instructions come in the form of genes, whose chemical composition contains specific information a cell uses to build proteins. Originally the gene of interest was recombined with the cell's own DNA. Today the gene is frequently carried on an independent bit of plasmid DNA, which the cell treats as its own. Second, the cell is encouraged to multiply in large vats in a process quite similar to brewing beer. At some point in the cell's growth cycle, it is induced to make the protein of interest. Finally, the protein is purified from the population of cells and becomes just another product in modern commerce.

These recombinant proteins, as they are called, are cropping up everywhere, including the pharmacist's counter. *Epoetin alpha* (also known as Epogen and Procrit) is a recombinant protein used to increase the production of oxygen-carrying red blood cells. Diabetics worldwide use recombinant human insulin. The antiarthritis drug Enbrel directly interferes with the molecular pathway that causes inflammation. These drugs are all derived from recombinant protein technology and produced in the manner described above.

The next steps in developing biological technology involve programming combinations of cells to do interesting things, such as produce valuable goods. This is already being undertaken in the case of insect larvae, plants, goats, and cows modified to produce drugs and useful proteins in their tissues and milk. Understanding and controlling consortia of cells, from novel networks of single cells to multicellular organisms, will open many interesting possibilities.

If you pause to consider it for a moment, we ourselves are sophisticated examples of just this sort of biological technology.

The human body consists of between 10^{13} (ten million million) and 10^{14} cells. That sum does not include the microbes living on and in our bodies, of which there about twenty times as many as our own cells. Some human cells are structural, some measure the environment, some process food into nutrients that other cells use, many communicate with each other within the body, and not one of them can survive without the rest.

Through this symbiosis, systems of human cells cooperate to produce the astonishing behaviors of speech, art, love, scientific inquiry, and religious devotion. If too many cells are damaged or otherwise begin to fail in significant numbers, all of the other interesting phenomena we can measure soon fade as well. So it is with the cells composing (virtually) every multicellular plant and animal.

Within the cells comprising organisms, intricate molecular systems process material and information. Some systems handle metabolism, taking care of the moment-to-moment processes necessary to sustain life. Other molecules maintain a cell's structural integrity or are involved in deciding how to respond to outside stimuli. This integrated molecular technology is ultimately controlled in the nucleus, where genetic instructions reside in DNA.

The elaboration of the function of all these molecules is inevitably obscured by a fog of jargon. Even though many names of molecules and processes are fairly straightforward, they often appear in complicated combinations that obscure simple details. For many readers the juxtaposition of familiar and unfamiliar makes the whole thing incomprehensible.

For example, outside of biology, the word "transcription" elicits thoughts of writing out or copying information from one format or medium to another. And this is precisely what transcription means in biology. Information stored in DNA must be transferred to a medium more suitable for handling and processing by cellular machinery, which in this case is messenger RNA (mRNA). The chemical code in the two media is very similar, and the molecular copy machine that transfers the information from one polymer to the other is a polymerase (figure 2.1).

Similarly, outside of biology the word "translation" conveys the notion of converting information from one language to another. Inside a cell, translation refers to the process of switching from the chemical code of mRNA, composed of nucleic acids, to the chemical code of proteins, composed of amino acids. The molecular machine responsible for translation is a ribosome.

Everyone understands what transcription and translation are in the context of normal usage in the English language, but add DNA, RNA, amino acids, polymerases, and ribosomes to a sentence and suddenly most people find it difficult to follow. It is a rare exposition that manages to convey details of molecular biology and chemistry while keeping the average reader's attention.

This difficulty exists in part because discovery of those details has been

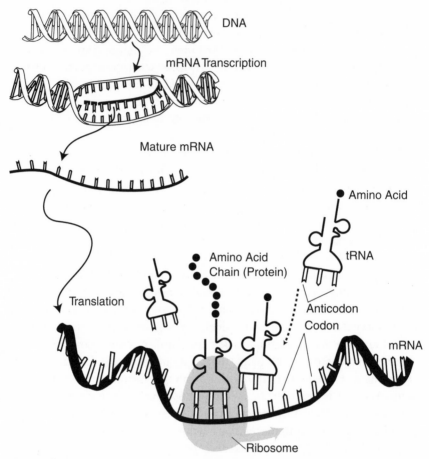

Figure 2.1 From DNA to proteins. The chemical code of DNA is first "transcribed" into the chemical code of mRNA by a polymerase (not shown). Then mRNA is "translated" into protein by a ribosome, which assembles amino acids into a polymer by matching transfer RNA (tRNA) to the code on mRNA. Adapted from the Talking Glossary of Genetic Terms, National Institutes of Health.

the most important story in biology for much of the last five decades. Those details literally *were* the story. Now, however, our understanding of many molecules is reaching the stage where we can describe them with a name that encapsulates a generalization of function. These generalizations also lead to the capability to mix and match molecular pieces to achieve a desired function. This is precisely the characteristic of automobile parts and computer components that enables much of modern engineering and manufacturing. The general functions of engines, carburetors, brakes, Pentium processors, and disk drives are widely understood without knowledge of the detailed engineering or fabrication that goes into those parts. That is, cars and computers are systems that can be composed from components of known input and output behavior.

Simple composable systems are familiar to many children, in the form of building blocks, which brings us back to LEGOS, Tinkertoys, Lincoln Logs, Erector Sets, and even wooden blocks. These toys are so intuitively used that pictures in instruction manuals are adequate to serve as predictive descriptions of real objects; you can be quite sure that by following the pictorial assembly instructions for complex LEGOS, the final product will look satisfyingly similar to the cartoon. The product will even possess the functions indicated in the cartoon.

With experience, a mental picture of a new LEGO artifact can be just as predictive as those cartoons, the mind's eye serving as a platform for creating structures through novel combinations of parts. It is not yet possible, in the vast majority of circumstances, to mix and match biological components with this same ease, because the behavior of each part is insufficiently defined.

It is the difference between complete composability and current reality that limits the construction of biological artifacts—molecular systems inside or outside cells, actual engineered cells, systems of cells, or ecosystems—with predictable behaviors. Of course, this does not mean that molecular components of biological systems will *never* be composable. Quite the contrary, it is the explicit goal of many researchers working at the intersection of biology and engineering to create composable molecular systems that function as predicted within cells. The promise of this effort lies in artificial biological systems that will allow us to produce food, chemicals, and other raw materials more cheaply and cleanly than we do now.

But there is something missing from this clean picture of engineering and manufacturing. It is even missing from the Honda Element commercial. The bricks that form the car come together of their own accord; there are no human hands involved in the assembly. In the real world, obviously, human hands are required for the time being at some level to put any object together. Those hands work at the behest of human brains (most of the time), and those brains bring along diverse personalities, varying levels of attention to detail, and all the many foibles that make us who we are.

Practical Considerations of the State of the Art

Science has brought us to the point where we are learning to control the molecular elements of life and the flow of information between them. Genetic modification is an alteration of function through the elimination or replacement of particular identifiable gene sequences. And when we find a new gene in nature that appears to be useful, we may be able to direct it to new purposes by moving it into foreign organisms.

This practice of manipulating biological systems toward a particular end is already contributing to new understanding, though the lessons learned are not always easy to take. The checkered history of gene therapy provides many such lessons. For example, it seems that a successful cure of a genetic disease has unintentionally caused cancer, in part because a design rule was not appreciated until after new construction was attempted.

Ten children in France born with X-linked severe combined immune deficiency (X-SCID), also called "bubble boy disease," were cured of that disease using gene therapy.[6] The experimental treatment consisted of adding to the children's genomes a functional copy of a gene critical to normal immune system function. Carrying genetically modified cells that make the functional protein, the children were part of a cadre of patients who were able to leave the confines of hospitals and lead mostly normal lives. Unfortunately, in some cases a new disease appears to have been caused by the successful gene therapy: two of these patients have since been afflicted with a peculiar leukemia.

The treatment for X-SCID involved removing blood stem cells (which produce red and white blood cells) from the patients' bone marrow, then genetically modifying the cells using a retrovirus. Retroviruses are capable of a remarkable bit of biological remodeling. They can introduce their own genetic material into the chromosomes of the cells that they infect, thereby ensuring the virus species' long-term survival. This naturally occurring engineering tool is often utilized in the laboratory to genetically modify cells by replacing most of a retrovirus's own genetic instructions with a different DNA sequence. The virus will then insert this sequence into the genome of the host cell, which results in a permanent change to the host genome. In the case of the X-SCID treatment, the retrovirus worked as planned, incorporating the therapeutic gene into the stem cell. In addition, genetic regulatory elements were inserted along with the new gene so that the cell would be more likely to make the new protein.

The modified cells were then injected back into the patients, where they began to reproduce as expected, churning out white blood cells containing the therapeutic gene. The gene seemed to function normally, producing protein at levels adequate to restore the patients' immune systems.

However, because this particular retrovirus inserts DNA into a random location in a stem cell's genome, there was a small chance that the inserted DNA would through its new position influence other genes. One potential explanation of the leukemias is that the inserted DNA "landed" in a location where it increased the activity of a nearby gene linked with cancer. A recent study in mice using the therapeutic gene found that the gene itself could cause leukemia because it alters signaling pathways that affect leukocyte development.[7] Whatever the ultimate cause in the human cases, exchanging one life-threatening disease for another was an unexpected consequence of the "cure."

The commingled ethical and practical challenges in the situation are obvious: children afflicted with X-SCID rarely survive into adulthood, but, now that a cure is available, giving X-SCID patients a potential for leukemia instead should not be part of the deal. Fortunately, the children have thus far responded to treatment for the leukemia, and through new techniques it should be possible to better control the location of gene insertions for this kind of therapy. The whole affair highlights the difficulties we face in trying to modify biological systems that we do not yet understand.

How Far We Have to Go

The X-SCID experiment was based on rules deciphered from observing natural systems and from initial attempts to edit control programs via gene therapy. The experience revealed a new design rule, namely, that with uncomfortably high probability the insertion of a therapeutic gene can induce unintended activity in nearby genes. Unfortunately, aside from notable exceptions that I will explore in future chapters, this short natural language statement—that is, a short story told using English grammar and vocabulary—is often as precise as predictive tools get in biology.

The limited predictability enabled by natural language descriptions is a clear impediment to efforts to reengineer complex systems, of which gene therapy is an example. The underlying problem is that we don't know all of the design rules for systems of interest. Moreover, we are trying to cross a frontier delineated not just by knowledge but also by the *quality* of knowledge. This is quality not in the sense of "good versus bad," but rather in the sense of qualitative versus quantitative, and whether the knowledge can be applied to understanding new situations or is limited to the original situation.

For example, the natural language design rule discovered during the X-SCID experiment is sufficient to avoid making the same mistake, but it is less useful for avoiding new mistakes. This gap makes explicit the step from science to technology. Accurately describing the behavior of unmodified cells does not require the new design rule. The only way to discover that rule is through tinkering. The questions of how and when scientists decide that they understand enough to try something new, especially when human lives are at stake, are clearly difficult. Good experiments always result in new knowledge. Yet will this new knowledge, once gained, reveal the choice to proceed as adequately informed or as sadly premature?

The question comes down to what we choose as the definition of "understanding." Does a natural language story of genes and proteins qualify? Does a design tool based on quantitative models qualify? How is this understanding manifested in our ability to interact with the real world? One definition is that the measure of understanding in biology should be the ability to build biological systems. That is, only when we can build a system that quantitatively behaves as predicted should we say we understand it. This definition of understanding sets an extremely high bar. It should probably be kept in mind as an ideal rather than the standard to be met before innovative, potentially life-saving treatments are tried. Yet it does place current levels of understanding in an interesting context because it illuminates our ignorance and the various risks we face.

Even if we accept the ability to build something as the *measure* of understanding, this still leaves the sense of the word hanging. For example, one criticism of this measure of understanding is that we might successfully parrot the function of biology without being able to explain why anything works; the measure does nothing to address the question of why anything is the way it is, in biology or any other subject.

With understanding so tenuously defined, we still face a decision of how to proceed in modifying biology. There are obvious benefits to be gained from improved biological engineering. There are also obvious hazards. How do we determine that we know enough to risk action, particularly when that means intervening in the course of a human life? Once a line is crossed, is it possible to step back? The choice of whether to let a child die from disease rather than take a chance on novel technology is difficult. What if we can decide these questions only through experiment? Yet these are exactly the same considerations that befall every application of a new technology, from hydroelectric dams to pharmaceuticals to baby formulas to car safety systems.

The conundrum we face is not due to the underlying genetic manipulations but rather human behavior. Biological technologies in human hands are value neutral—neither intrinsically good nor bad—because the technologies that humans use are often adopted or adapted from nature rather than invented. The act of moving genes around is demonstrably neither intrinsically beneficial nor intrinsically dangerous, but any particular manipulation of the natural world involving human hands has the potential to be risky.

The larger issue is that, collectively, we have already made decisions that are not reversible. As I will argue in later chapters, given the state of technology development, it is simply too late to go back. Nor is it obvious that we could have avoided our current situation. In any event, while it can be tempting to guess at what the world may have been like if decisions had been made differently, we must deal with the world we live in today. Human curiosity and ingenuity have provided the ability to modify biological systems but not yet the ability to rapidly recognize missteps or to rectify the products of mischief or mistake. We must address these gaps as quickly as possible.

Biology is a young technology in the hands of humans. By facing up to how little proficiency we now have and, by proceeding accordingly, we can develop more-realistic expectations of both the current state of biological technologies and the dramatic possibilities for the future.

CHAPTER THREE

Learning to Fly
(or Yeast, Geese, and 747s)

T HERE IS AN OLD SAYING in physics that geese can't fly. It is a mystery to this day how geese manage to carry all that mass around on long flights. Our understanding of the physics of flight suggests that geese should not be graceful and efficient, despite clear evidence to the contrary. Estimates of how much power is required for a bird to maintain a certain mass aloft have always been somewhat confused, as illustrated by the difference between previous theoretical predictions and experimental measurements published in the journal *Nature* in 2001.[1] It turns out that heavy birds are considerably more metabolically efficient than we have given them credit for. The lesson is that we still have a long way to go in explaining even seemingly simple biological phenomena that are plentiful just outside the window. This may seem a bit incongruous, given the fact that we build aircraft that fly considerably farther and faster than the average honking goose.

In the same issue of *Nature,* another paper made some progress in resolving the ongoing debate about whether birds fly in formation for physical or social reasons. It turns out that a bird flying in formation can ride updrafts created by the bird just ahead, allowing more gliding and requiring fewer power strokes.[2] A theoretical prediction that flying in formation is energetically beneficial was made in 1914, but it took another eighty-five years to make the measurement. Even so, the paper claims only qualitative, rather than quantitative, agreement with theory. This means that experiments can now be done with sufficient precision to confirm that flying in formation is generally a good idea for birds, but the experiments are not yet good enough to confirm detailed numerical predictions of the theory. It may still turn out that the theory gets the details wrong and must be revised.

The discrepancy between theory and experiment is a characteristic of the sorts of stories we tell about the way biology works: It is much easier to write down a quantitative description of a biological system than it is to test that description. The primary division between the study of biological systems and that of physical systems more amenable to traditional engineering can be found in storytelling. Most models and experimental predictions in biology are natural language stories. Models of protein function today often have the structure of "Protein X binds to Protein Y" or "Protein X recognizes and cleaves a certain DNA sequence." For the most part, there aren't any numbers included. These sorts of models lack quantitative predictive power, whereas engineering generally requires a framework of quantitative models based on quantitative experiments. Quantitative experiments have certainly been done in biology, and models constructed to describe the results, but they have generally been aimed at describing the behavior of populations rather than individuals.

Indeed, there is a long and fruitful history of experiments and models dealing with the behavior of a system of molecules or a system of organisms. It is, in many cases, relatively straightforward to create statistical descriptions at the level of populations. This could be a population of molecules, for which quantitative tools in statistical mechanics and biochemistry have been very successful, particularly in describing the many molecules that contribute to the electrical behavior of neural cells. Statistical descriptions also accurately predict the way bacteria and other single-celled organisms respond to chemical gradients, known as chemotaxis. Alternatively, statistical descriptions can be created for a population of rabbits (prey) and foxes (predators). Ecological modeling derived from MacArthur and Wilson's seminal work, *The Theory of Island Biogeography,* is both quantitative and predictive.[3] The existence of quantitative population models in these fields provides a basis for biochemical engineering and the beginnings of ecological engineering, respectively, which both deal with the statistical behavior of large numbers of individuals. It does not, however, help predict the behavior of individual members of the population nor of specific species in real ecosystems, whether genetically modified or natural.

This difference is particularly important when considering the transition of a field of study from "pure" science to technology. We are extremely interested, for example, in how much water a *particular* dam can hold back, rather than in a qualitative story about how much water can be restrained by a pile of concrete or even how much water dams hold back on average. Thus our limited understanding of biological aviation betrays something profound about our ability to implement real biological technology. While technological progress based on natural language stories is certainly possible, quantitatively predictive models, in contrast, rapidly become design tools and enable true engineering.

The history of human attempts at heavier-than-air flight demonstrates this point very well. Early investigations into flight were disparate endeavors guided as much by ego and opinion as observation. In the absence of a theoretical underpinning, it took many years of effort to lay stable foundations for human-powered flight. But soon after controlled flight was finally demonstrated, aviation became a discipline in its own right and soon changed the world.

The Frenchman Louis-Pierre Mouillard, in his 1881 book *L'Empire de l'Air,* suggested that powered flight would only be possible after learning how to control airplanes via gliding. He distinguished "pilots," those possessed of skills that would enable them to fly aircraft, from "chauffeurs," who were focused on powering their way into the air without any knowledge or ability to control their craft.[4] Mouillard was truly ahead of his time. In a letter to Octave Chanute in 1891, he identified aluminum as "the metal for aviation."[5] Despite Mouillard's insight into the need for flight controls and new materials, as an early aviation pioneer he managed only short flights with small model airplanes and never produced a truly functional passenger-carrying glider.[6]

The approach to aviation elucidated by Mouillard provides useful insight into the future of biological technologies. It is only through learning how the parts work, and how they work together, that we will truly be able to produce engineered biological systems. Just as we learned to fly aircraft, so must we learn to "fly" biology.

From Stories to Data . . .

Aviation emerged in a distributed fashion under the influence of only the most embryonic notions about general principles of flight. Early attempts at building artificial wings were in some cases based on philosophy as much as science. Otto Lilienthal made his first glider flight in Germany in 1891 and later spent several years making vaguely controlled jumps off an artificial hill near Berlin. He was among the first early experimenters to recognize that cambered, rather than flat, wings were required for flight. He was also among the first to implement Mouillard's notions of control, flying his craft by shifting his weight around the center of lift, foreshadow-

ing the flight of today's hang gliders. But Lilienthal was cursed by an ancient Greek preference for geometric forms, which led him to favor airfoils
that were sections of perfect circles.[7] At first glance, the curve of a bird's
wing may indeed look like a section of a circle, but evolution spent many
years discovering that parabolic arcs are the better shape for flying.

While Lilienthal struggled with an airfoil shape that was inefficient and
thus difficult to control, Octave Chanute spent the better part of the 1890s
compiling data that suggested a parabolic arc was indeed superior. Chanute
may have learned of the shape from John Joseph Montgomery, who is
credited with the first use of parabolic wing sections on a glider in 1883.[8]
Near the shores of Lake Michigan in Indiana, Chanute's assistants and collaborators flew gliders with parabolic cross-section wings several hundred
times in the late 1890s.[9]

Chanute was generous with his data. Through correspondence with all
the major experimenters of the day, he freely gave away his findings, including extensive data tables on airfoil efficiency, providing a core of airfoil design information to the community.[10] The Wright Brothers were
recipients of this generosity and benefited from Chanute's enthusiasm, encouragement, and design for the now familiar braced-box (also called a
Pratt truss) biplane construction. They put the latter information to good
use, combining it with their own extensive observations and experimental
results, first taking wing in a glider in 1900. Experience in flying this 1900
glider gave the Wrights enough confidence to later affix an engine to a craft
of very similar design, making the first powered flight by a human in December of 1903.[11]

Although Chanute and the Wrights were frequent correspondents in the
1890s and early 1900s, Chanute temporarily fell out with the Wrights during their efforts to patent aviation technology. He felt that the information
and inventions should be a public good, that the Wrights were impeding
broader development, and that he deserved some credit for their specific
inventions. The parties had begun to reconcile at the time of Chanute's
death in 1910.[12]

The program of careful observation and methodical experimentation
begun by Chanute and the Wrights was slowly augmented by a growing
body of empirical evidence concerning the flow of air over and around various shapes. But during its origins, the early practitioners of aviation relied
on shared, though differing, stories, including Chanute's data tables describing the efficiencies of different airfoil shapes. Only with time were
conclusions based on this data justified with physics. In particular, Lilienthal's mistaken assumption about airfoil shapes contributed to significant
control problems in his aircraft, leading to a serious crash in 1896. Lilienthal died from his injuries. The Wrights diagnosed the cause of his crash

several years after it occurred, leading, in part, to the design of the novel control features in their own glider.[13]

. . . to Design

Throughout this early age of discovery and invention, and in the decades that followed, the beginnings of a quantitative theory of flight encompassing airfoils, control systems, and power systems slowly accumulated in the minds of mathematicians, physicists, and budding aeronautical engineers. Thus the early qualitative stories about airfoil shape and wing construction evolved into a body of knowledge that superseded individuals and egos, becoming codified in more mathematical terms. Theory and experiment advanced in turn. Hypotheses were confirmed or overturned by experiment, and experiments were explained and guided by new hypotheses.

At the end of this process—if it is the end—aeronautical engineering has advanced to the point where theory is so accurate that airplanes are designed on computers, tested on computers, and built predominantly without wind tunnel testing. The first airframe of a new model is often flown as soon as it is constructed. The Boeing 777 was the first airplane built in this way. It was designed based on expertise from building previous aircraft and fifty years of intense academic and industrial effort compiling experience, theory, and shared technologies.

It is remarkable that development of aviation technological has recently returned to where it started: the garage. The X Prize was intended to inspire the same ferment in spacecraft as existed in the early days of powered flight, and it succeeded admirably. As demonstrated in 2003 by Burt Rutan and his crew at Scaled Composites, building a spacecraft is now within the range of relatively modest investments, although not every inventor has $25 million to spend on garage projects. Even though Rutan's SpaceShipOne was not capable of reaching orbit during its demonstration flights, until recently building any sort of functional spacecraft was beyond the reach of everyone save a few governments and large corporations. SpaceShipOne was designed and tested primarily on inexpensive desktop workstations running commercial computational fluid dynamics (also known as CFD) software. When physical testing was necessary, the builders strapped the ship to the back of a pickup truck and drove through the desert at high speed. The success of the project owed as much to the expertise and experience that produced the 777 as it did to Rutan's brilliance as a designer. Improvements in physical, conceptual, and computational tools are what made this possible.

The history of aviation, with its progression from qualitative stories to computer-based design, very likely reveals the future course of biological technologies. But before we get overly excited over the possibility of rational biological design, it is instructive to compare the state of our knowledge of biology today with that of our knowledge of aviation technology. In particular, we can examine how well we know an organism we have relied on for thousands of years.

Yeasts are single-celled organisms that enable us to produce beer, bread, cheese, and wine and may well serve as an initial manufacturing platform for other biologically constructed items. How does our knowledge of the bits and pieces that make up yeast compare to our knowledge of the parts in a workhorse aircraft like the Boeing 747?

The original design and construction of the 747 represents roughly the midpoint of the journey from the Wrights to the 777. The 747 is built from approximately 50,000 kinds of parts, some of which appear many times (e.g., fasteners), such that there are 6 million total components in a 747. The aircraft is built on 50-year-old technology. Each part, whether wing strut, flap, or turbine blade, is described by its own quantitative model—its *device physics*—that predicts its behavior under the conditions expected during flight. The notion of device physics is the cartoon at the heart of composability (introduced in Chapter 2). It is a highly functional abstraction that allows design using relevant features of a component as a whole rather than, for example, focusing on what all the atoms in a turbopump are doing during takeoff.

All of the structural shapes in a 747 are the product of long experience optimizing weight and performance, and the materials chosen for each part are thoroughly tested for strength and longevity. The description of each piece of a 747 fits within a larger model that originally resided partially in blueprints and partially in the heads of Boeing engineers. Each part was tested to ensure that it behaved as specified by the overall model. Each part met those design specifications.

What about yeast? Beginning with only the components described by their genome (proteins and RNAs), yeast have approximately 6,300 kinds of genetic parts and are built with technology that is 3 billion years old. We have models for about 50 percent of those parts, and those models are built at very low resolution. That is, we have given each of those parts a name, we know generally what sort of molecule each is, and we may have some knowledge of its function. We do not know most of the design specifications of the parts (i.e., what they are for, and why), and we know the quantitative device physics for only a handful. There are between tens and thousands of copies of most proteins active in a yeast cell at any given time.

Including the molecules in yeast not directly described by the genome would swell the parts list by many more thousands, and only some of those molecules are known in name and in function. (In the very limited description I employ here, genes encode proteins—there is no genetic encoding of the structure of a sugar or a lipid. These molecules are constructed or manipulated by proteins.) These molecules include the lipid and carbohydrate components of the cell membrane, and the various components of metabolism. This list is continually growing, and we have only a rough idea of how many of each part are present in the cell, which other parts they interact with, and under what conditions. The total number of moving parts in any given yeast cell, while still unknown, certainly amounts to many, many millions.

One important difference between designing biological systems and designing airplanes is that with airplanes we choose to work within performance envelopes that are relatively easy to model and understand. Clearly the range of possible flying things exceeds those built by humans. It is just as clear that we do not understand the physics of most of the things in the air, as demonstrated by the difficulty in explaining how geese stay aloft. But by restricting ourselves to modes of propulsion and shapes that are within our modeling skills, we can base design tools on those models and thereby provide an infrastructure for building airplanes like the 777.

A considerable technological advance from the 747, the 777 is built from approximately 130,000 kinds of parts (with about 4 million components in each airplane) and was designed from scratch entirely using computer-aided design (CAD). The engines add around another 50,000 parts, some of which move at supersonic speeds. During the design phase, the device physics and the design specifications for each part resided in a large computer model that was repeatedly simulated in order to understand the likely operation of the completed aircraft. Though the individual parts and assembled subsystems, such as doors, were thoroughly tested, there were no engineering mock-ups used in construction of the aircraft.[14] The first 777 airframe served as a flight test bed. Boeing put the airplane together from its constituent parts and then flew it straight away. It was a remarkable engineering feat.

We trust our lives to this process every day.

In contrast, we have very few quantitative models of biology, and we have no design tools for biology. Despite the paucity of such tools, we forge ahead (with mixed success), building rudimentary synthetic biological systems, introducing genetic modifications into the environment.

We trust our lives to this process every day.

The difference between the two technology sets is as striking as our trust in both. Mathematical models of the behavior of all the systems on passenger jets, and of the airframe itself, are so well tested and trusted—I will call this the "flight model"—that many heavy-weather landings involve no human control whatsoever. Models of an airplane's performance built during development become the basis for autopilot programs. Because today's flight models are so good, autopilots based on them can take care of all situations encountered in rain, fog, and high winds. Almost everyone who has flown on a commercial jet in the last few years has been the beneficiary of such abilities, including takeoff and landing.[15]

The heart of this capability, of course, resides in onboard electronics, also known as avionics. In short, the combination of computers and quantitatively predictive models of airplane behavior provide for very impressive technology. But there are distinct differences between the flight model, the algorithms used to simulate the model, and the computers on which those simulations run. The existence of fast computers alone does not provide a functional, safe airplane design, nor are those computers sufficient to fly the resultant craft.

While predictive models describe the behavior of airplanes under a broad range of conditions, the most important feature of flight is the interaction of the craft and its environment. For example, while every effort is made to keep turbulence near wings and control surfaces to a minimum in normal flight, the flow of air over surfaces is extraordinarily complex and in general is very difficult to predict. A sudden transition from laminar, or smooth, airflow to turbulent airflow over control surfaces can dramatically affect the stability of an airplane. Turbulence is a clear example of a situation when events outside the control of the autopilot—and outside the predictive capabilities of the flight model—can produce unexpected consequences.

The discrepancy between what the autopilot expects to happen based on the flight model and what actually happens at each moment is resolved through very fast and accurate measurement of the behavior of the airplane. This process produces information that the autopilot can use to alter the position of control surfaces and thereby correct deviations from what the flight model predicts.

In a large and stable aircraft like the 777, this feedback is used primarily to reduce the impact of uncertain environmental conditions. But flight models are also now accurate enough that computers can fly airplanes designed to be so unstable that humans cannot pilot them unassisted by automation. The X-29, an experimental aircraft built in the early 1980s, was designed to take advantage of the instability resulting from its forward-swept wings and consequently was more maneuverable than any airplane of the day. But controlling the radically designed aircraft required subtle manipulation of wing shape and constant monitoring of aircraft attitude. Humans simply could not assimilate and process information quickly enough to keep the plane in the air.

Computer speeds in the early 1980s were just fast enough to handle such tasks, provided you had access to NASA and Department of Defense bank accounts to build prototypes. Now, of course, processors run at speeds orders of magnitude higher. However, technological progress that increases computing speed does not change the fundamental interaction of measurement, model, and computation required to fly airplanes. Improvements in processor speed and the underlying algorithms used to simulate and explore the behavior of the flight model certainly endow flight models of today's commercial airliners with greater capability than ever before. But the expectation of the model must still be checked against the real world, even though most airliners in the sky today have vastly more capable computers onboard than did the X-29.

It is conceivable that, had computational resources been available before flight was developed, theory and simulation might have preceded experimentation and the resultant copious data on airfoil efficiency. Purely theoretical insights occur from time to time in physics, such as Einstein's invention of the theory of special relativity or Paul Dirac's prediction of the existence of antimatter. But the origin of human aviation at the turn of the twentieth century, like the vast majority of scientific and technological problems, was driven by experimental failure and success, followed by a need to explain and understand. The development of biological technologies will very likely follow a similar course, despite enthusiastic claims to the contrary.

Both the popular and scientific press today repeat the notion that accurate, predictive simulations of biological systems are soon to be in the offing and that fast computers are the reason why. These promises are not solely hype. Behind them is sincere hope that great theoretical insights in biology are imminent. And computers will without a doubt play an important role in the development of biological technologies. Much has been made of their role both in unraveling biological mysteries and in developing the engineering that will follow. But computational resources by themselves are not adequate to produce new understanding. Computers are of great help in handling large amounts of data, but in molecular biology— unlike physics, chemistry, and most engineering disciplines—purely theoretical insight has been rare. This trend is likely to continue until better experimental measurements support the construction of predictive models and better techniques are developed to simulate the models.

Building and Simulating Models
of Biological Systems

The difference between model and simulation deserves more attention, for it is important in some challenges that lie ahead. A model is an embodiment of knowledge about a system, consisting of, for example, qualitative stories, empirical descriptions, or equations. In the field of physics, the "Standard Model" of particles and forces is composed of equations that describe the interactions between particles and of experimental data that constrains the masses of those particles. A simulation of the Standard Model is an attempt to calculate something interesting—say, the amount of radiation emitted in a nuclear explosion or the behavior of a photon as it propagates through a new material—in the service of making a prediction about the results of an experiment. The Standard Model has been simulated in many different ways to make predictions about the behavior of fundamental particles in different conditions and arrangements, and those simulations have often been used to identify experimental conditions that might reveal new particles. Improvements in all these areas are necessary for progress in physics or any other science—Nobel Prizes have been awarded for the construction of new pieces of the Standard Model, for calculations that predict the existence of particles or connect pieces of the Model, and for experimental confirmation (and sometimes refutation) of specific predictions. The utility of a model in making predictions is often determined by the kind of simulation used to make calculation, and as computers and computer science have progressed, so has the range of conditions that can be simulated using the Standard Model. Calculating—simulating—the mass of a particle or the results of a stellar collision is a very different experience when using a pen and paper than when using a supercomputer. Thus the ability to successfully simulate nature is dependent both upon the means used to run the simulation and upon the structure and contents of the model.

In this section, I will concentrate on what it means to build and test models of biochemical and genetic networks. I will completely ignore the important fact that spatial variation and compartmentalization within a cell appear to be critical design features. While the physics of diffusion and transport within cells are being studied quantitatively, it is difficult enough to build a model of the network without also worrying about what is happening where within a cell. Clearly, real design tools for biological systems will need to include all this information.

The 747 flight model is a synthesis of the device physics of all the components in the airplane. More than a collection of facts, the structure of a model defines the relationships between variables included in that model. The present structure of most biological models is determined by English grammar because the models are constructed using natural language, as described at the beginning of this chapter: "Protein X binds to Protein Y" or "Protein Z is implicated in many breast cancers." In the future, as quantitative knowledge improves, models of biological systems will contain increasing amounts of algebraic structure that is more amenable to computational simulations. For example, "The concentration of complex XY is proportional to both the concentration of Protein X and the concentration of Protein Y" is easily expressed in a simple equation: $[XY] = (1/K)[X][Y]$. The constant of proportionality, called the dissociation constant in this case and represented by K, describes how strongly Protein X binds to Protein Y. By definition, a small value for K means that most of the reactants, X and Y, are bound together.

A single chemical reaction of this sort is fairly easy to understand and to juggle mentally, but an alphabet full of proteins that interact in multiple ways presents considerable difficulty, in part because there is a different K for each interaction. It is straightforward to write down a model for such an alphabet soup, but simulating it is another matter. The lack of knowledge of those dissociation constants—as well as of the rate constants that characterize how fast each reaction occurs—is what, in most cases, currently makes simulation difficult.

In cases where numerical simulation of chemical reactions is possible, it is often used to explore the behavior of a model with particular values for the variables appearing in the model. If a dissociation or rate constant is unknown, simulations can be used to explore what happens to the model with a wide variety of possible values.

Even if all the values of relevant constants are known, the ability to adequately simulate a model to make accurate predictions is in general distinct from the formulation of the model. Simulation could take place with pencil and paper, in the case of just Protein X and Protein Y, or it could require small supercomputers to simulate all the proteins and other key molecules in a bacterium, an effort under way in multiple research centers around the world.

The line between model and simulation becomes blurred when simulation is the only way to explore a particular model or when predictions from the model depend on which simulation method is used, as described in the next chapter. This difficulty underlies many simulation efforts in physics, chemistry, and engineering. For example, when simulating the flight model of an airborne 747 full of passengers, the autopilot clearly must use a simulation method that enables accurate predictions of the behavior of the airplane. Fortunately, all those details are checked out in the design stage of the airplane, before the autopilot is trusted with a human life. Similarly, when working on projects closer to invention than product, scientists and engineers spend a great deal of effort ensuring that simulations faithfully represent the real world. As with the airborne 747 on autopilot, this is accomplished by repeatedly comparing simulation and measurement through experiments. Before the simulation can be used for predictions, it must be as trustworthy as the experiment. It would not do to rely on a simulation that was inadequately tested.

Despite an obvious need for clear distinction between results of simulations and the behavior of objects in the real world, some researchers refer to simulations themselves as "experiments." This may be inspired by an expectation that the simulation accurately represents phenomena in the real world or perhaps by the ability to run multiple simulations to explore the behavior of a model over a range of initial conditions.

While it is probably not particularly good terminology under most circumstances to call the results of a simulation an "experiment," there are in truth a few instances where it is acceptable. A flight model is one such instance—it is clear that simulating the flight model of a 747 produces extremely accurate predictions of a physical system. Simulation of nuclear weapon detonation is arguably the best example, in that the United States has for many years built new nuclear weapons without actually testing the new designs. Biology is not yet on the list. There is nothing in biology remotely approximating a flight model.

To be sure, biochemists have met with considerable success in quantitatively describing chemical reactions utilized in biological systems. Textbooks are full of such examples.[16] The basic theory describing most biochemical reactions has been thoroughly vetted by experiment, and there is no reason to think any of it wrong (though it may, of course, be incomplete). But while that theory provides a basis for describing the behavior of molecules, it also provides considerable leeway. Without very careful measurements of molecular properties, it is difficult to take the general behavioral description provided by theory and produce quantitative predictions; without knowing the dissociation constant, it is nigh impossible to predict the concentration of XY. Only when adequate measurement can be made of, say, how strongly one protein binds to another, or the rate at which an enzyme processes its substrate, can numbers be added to natural language predictions. The absence of this kind of data is a considerable impediment to a deep understanding of biological systems.

For example, predicting the influence of a new drug on human metabolism would be a great leap forward in drug development and health care.

A quantitatively predictive model of metabolism could potentially aid in avoiding serious side effects that are currently discovered only through trial and error. Unfortunately, current knowledge is generally insufficient for such simulation. While a small number of the individual interactions in human metabolism have been quantified, the majority remain vague, and we rely on natural language models for understanding. The best summaries of metabolic networks to date are charts containing names of molecules and arrows indicating interactions.[17] The charts contain no numbers and are therefore incomplete descriptions of the phenomena. Not only is quantitative information crucial for understanding individual reactions, it is absolutely required for understanding *networks* of interactions, such those that constitute metabolism. As we will see in a concrete example in the next chapter, even a simple network can display *qualitatively* different behavior—for example, oscillation instead of static behavior—depending on *quantitative* features of its constituent molecules, such as dissociation constants, rate constants, and molecular lifetimes. It is certainly possible to explore these different modes of behavior by simulating models of reaction networks and plugging in good guesses for the requisite, but unknown, parameters. But those simulations often display behavior that is qualitatively different than that which is observed in experiments, revealing that the models do not contain adequate details pertaining to real systems.

An alternative to the struggle inherent in building models of existing complex biological systems is to choose systems that are easy to model. This is a general principle of human manufacturing, from cars to computers. And for most of human experience with aviation, for example, we did not attempt to build aircraft that flew like insects or birds. Now, however, measurement, modeling, and computation have improved to the point where principles divined from studying insects are being used to design small robotic aircraft. Just as we choose to build flying machines that operate within our ability to model, so can we begin building synthetic biological systems that operate within our ability to model, using design specifications based on the few individual parts for which the device physics exists. This approach produced the car you drive, the computer on your desk, and, if you have purchased the latest Nike or Adidas trainers, probably the shoes on your feet. The next chapter examines recent efforts to build biological systems "to spec," with examples chosen to give a sense of how biology is becoming quantitative and what challenges must be overcome along the way.

CHAPTER FOUR

The Second Coming
of Synthetic Biology

"I MUST TELL YOU that I can prepare urea without requiring a kidney of an animal, either man or dog."[1] With these words, in 1828 Friedrich Wöhler claimed he had irreversibly changed the world. In a letter to his former teacher Joens Jacob Berzelius, Wöhler wrote that he had witnessed "the great tragedy of science, the slaying of a beautiful hypothesis by an ugly fact." The beautiful idea to which he referred was vitalism, the notion that organic matter, exemplified in this case by urea, was animated and created by a vital force and that it could not be synthesized from inorganic components. The ugly fact was a dish of urea crystals on his laboratory bench, produced by heating inorganic salts. Thus, many textbooks announce, was born the field of synthetic organic chemistry.

As is often the case, however, events were somewhat more complicated than the textbook story. Wöhler had used salts prepared from tannery wastes, which adherents to vitalism claimed contaminated his reaction with a vital component.[2] Wöhler's achievement took many years to permeate the mind-set of the day, and nearly two decades passed before a student of his, Hermann Kolbe, first used the word "synthesis" in a paper to describe a set of reactions that produced acetic acid from its inorganic elements.[3]

At the dawn of the nineteenth century, chemistry was in revolution right along with the rest of the Western world. The study of chemical transformation, then still known as alchemy, was undergoing systematic quantification. Rather than relying on vague and mysterious incantations, scientists such as Antoine Lavoisier wanted to create what historians sometimes refer to as an "objective vocabulary" for chemistry. Through careful measurements made by many individuals, a set of clear rules governing the synthesis of *inorganic,* nonliving materials gradually emerged.

In contrast, in the early 1800s the study of *organic* molecules was primarily concerned with understanding how molecules already in existence were put together. It was a study of chemical compositions and reactions. Unlike the broader field of chemistry taking shape from alchemy, making new organic things was of lesser concern because many thought that organic molecules were beyond synthesis. Then, in 1828, Wöhler synthesized urea. Over the following decades, the growing ability to assemble organic molecules from inorganic components altered the way people viewed the natural world, because they could now conceive of assembling complexity from simpler pieces. Building something from scratch or modifying an existing system requires understanding more details about the system than simply describing how it behaves. This new approach to chemistry helped open the door to the world we live in today. Products of synthetic organic chemistry dominate our manufactured environment, and the design of those products is possible only because understanding the process of assembling molecules revealed new principles.

The step of codifying synthetic chemistry, and then chemical engineering, radically changed the way people understood chemistry. Through construction, chemists learned rules that had not been apparent before. In the same way that chemical engineering has changed our understanding of nature, as we begin engineering biological systems, we will learn considerably more about the way biological pieces work together. Challenges will arise that aren't obvious from observations of whole, preexisting, functional systems. With time, we will understand and address those challenges, and our use of biology will change dramatically in the process. The analogy at this point should be clear: we are at the beginning of the development of synthetic biology.

Before going further, it is worth noting that the phrase "synthetic biology" has been used previously. The first time the phrase was introduced it was a flop. In her history of the modern science of biology, *Making Sense of Life,* Evelyn Fox Keller recounts efforts at the beginning of the twentieth century to discover the secret of life through construction of artificial and synthetic living systems: "The [path] seemed obvious: the question of what life is was to be answered not by induction but by production, not by analysis but by synthesis."[4] This offshoot of experimental biology reached its pinnacle, or nadir, depending on your point of view, in attempts by Stéphane Leduc to assemble purely physical and chemical systems that demonstrated behaviors reminiscent of biology. As part of his program to demonstrate "the essential character of the living being" at both the subcellular and the cellular level, Leduc constructed chemical systems that he claimed displayed cell division, growth, development, and even cellular mobility.[5] He described these patterns and forms in terms of the well-understood physical phenomena of diffusion and osmotic pressure. It is important to note that these efforts to synthesize lifelike forms relied as much upon experiment as upon theory developed to describe the physics and chemistry relevant to his synthetic systems. That is, Leduc tried to follow a specific program using physical principles to explain biological phenomena. These efforts were described in a review paper at the time as "La Biologie synthétique."[6]

While the initial reception to this work was somewhat favorable, Leduc's grandiose claims about the implications of his work, and a growing general appreciation for complicated biological mechanisms determined through experiments with living systems, led to a backlash against the approach of understanding biology through construction. By 1913 one reviewer wrote, "The interpretations of M. Leduc are so fantastic . . . that it is impossible to take them seriously."[7] Keller chronicles the episode within the broader historical debate over the role of construction and theory in biology. The scientists in the synthetic camp, and in related efforts to build mathematical descriptions of biology, were poorly regarded by their peers. Perhaps inspired by contemporaneous advances in physics, the mathematical biologists and the synthetic biologists of the early 1900s apparently pushed the interpretation of their work further than was warranted by available data.

In response to what he viewed as theory run rampant, Charles Davenport suggested in 1934, "What we require at the present time is more measurement and less theory . . . There is an unfortunate confusion at the present time between quantitative biology and bio-mathematics . . . Until quantitative measurement has provided us with more facts of biology, I prefer the former science to the latter."[8] I suggest that these remarks are still appropriate today. Leduc, and the approach he espoused, failed because real biological parts are more complex, and obey different rules, than his simple chemical systems, however beautiful they were.

Eighty years later, the world looks very different. Mathematical approaches are flourishing in biology, particularly in the interpretation of large data sets produced by studies of all the many genes and proteins found in an organism. But I think it is important to acknowledge that not all biologists agree a synthetic approach will yield truths applicable to complex systems that have evolved over billions of years. Such concerns are not without merit, because as the quotation from Charles Davenport suggests, biology has traditionally had more success when driven by good data rather than by theory.

Modern Synthetic Biology

The modern reincarnation of the effort to build living systems has many faces in many locations around the world. Scientists are at work to expand upon terrestrial biochemistry, changing not just the sequence but also the content of the genetic code beyond four base pairs. Other efforts aim to introduce into living systems new amino acids not found in any living organism. Through engineering proteins and genetic circuits, synthetic biology is in full ferment. The phrase "synthetic biology" itself has slowly reemerged in scientific and popular literature.

The earliest modern usage of the phrase appears in a 1974 article by Waclaw Szybalski that reviewed advances in understanding transcription and translation: "Up to now we are working on the descriptive phase of molecular biology . . . But the real challenge will start when we enter the synthetic biology phase of research in our field. We will then devise new control elements and add these new modules to the existing genomes or build up wholly new genomes."[9]

Szybalski invoked the phrase again in an editorial celebrating the 1978 Nobel Prize in Physiology or Medicine, awarded for the discovery of enzymes that enable recombinant DNA technology.[10] Splicing a gene from one organism into an unrelated organism is a step loosely analogous to Wöhler's synthesis of an organic molecule: it is the fundamental capability required to build synthetic biological systems and thereby begin determining design rules. It has been clear for many decades now, at least to some, which way biology has been headed.

A more-recent review categorizes efforts in the nascent field as follows:

> Synthetic biologists come in two broad classes. One uses unnatural molecules to reproduce emergent behaviours from natural biology, with the goal of creating artificial life. The other seeks interchangeable parts from natural biology to assemble into systems that function unnaturally. Either way, a synthetic goal forces scientists to cross uncharted ground to encounter and solve problems that are not easily encountered through analysis.[11]

Again we confront the idea that building new things forces us to acknowledge and solve problems that are not apparent simply through studying life as we find it. This is true whether one is trying to rebuild biochemistry from the ground up or trying to reconfigure the arrangement and operation of existing genes to enable new functions. For reasons that I hope will become clear as the book progresses, the goal of building synthetic biological systems from "interchangeable parts," despite being both audacious and exceptionally challenging, is likely to provide results sooner than many realize.

The most explicit and nuts-and-bolts efforts at pursuing what is sometimes called "the parts agenda" were initially led by a group at the Massachusetts Intitute of Technology. From the SyntheticBiology.org website: "Synthetic biology refers to both (a) the design and fabrication of biological components and systems that do not already exist in the natural world and (b) the re-design and fabrication of existing biological systems."

After exploring the development of aviation technology in the last chapter, it is natural to ask what it will take to build biological systems "to spec" in the same way we now build airplanes. We have seen that the differences between the two kinds of systems are enormous. Recall that there are many more moving parts in a cell than in an airplane, and we simply don't yet know how to fly a cell. As with early aviation, biological technology is in the process of moving from qualitative stories to quantitative models.

The design of airplanes is a highly mature field compared with the design of biological systems. There are also fundamental differences in the components immediately available for use. Unlike early aviation, for which all the components were fabricated from scratch, building most new biological systems currently requires the use of molecular components that have been evolving for several billion years. The main difficulty caused by this requirement is that we don't yet know how all those components work. One reason for such ignorance is that the physics governing the behavior of molecules is different from the physics we are used to at much larger length scales. The mechanical intuition we have for gears, motors, beams, and cantilevers cannot be simply transferred to molecules. The molecular scale of the components also means they are very hard to interact with directly, which complicates testing models through experiments. Fundamental aspects of basic components used to turn genes on and off, and to make proteins, are still mysterious. As explored in Chapter 2, designing biological systems just isn't like imagining a new LEGO structure. But even given our inadequate knowledge of biological details, a design philosophy based on composable parts still has utility. This is where cartoons, based on generalizations and abstractions of molecular details, can be of assistance.

Cartoons are used in chemistry to describe the states of atoms and molecules and, if you know how to read them, can help predict the outcomes of chemical reactions. To push this further, in high-energy physics, diagrammatical systems have been developed that allow interactions between particles to be predictively, and conveniently, visualized and that can also be used to calculate. Feynman diagrams, initially developed by Richard Feynman, are cartoons that have strict mathematical interpretations. There is an algebra—a grammar—governing manipulation of the diagrams and their translation into mathematics.

Cartoons are already playing a role in understanding biology, whether they illustrate molecular interactions or the information content of genes.[12] With a reasonable knowledge of genetics and biochemistry, gene and protein sequences can be read as if they were English. Chemical structures can help us figure out how two biological molecules will interact. Where phenomena are not directly observable via microscopy, cartoons are often used to suggest models of cellular organization and function.

Cartoons do not always make science easier. At the cartoon level it is much more complicated to predict the outcomes of assembling new biological structures, be they molecules or organisms, than it is to envision what a new LEGO construction will look like. In addition to there being many more kinds of biological LEGOS than there are plastic bricks, the biological building bricks tend to change their shape and color every time two bits are snapped together.

Toward a "Standard Model" of Biology

The first *experimental* evidence connecting the storage of heritable genetic information with DNA came in 1944, in the form of a natural language cartoon. Oswald Avery, Colin MacLeod, and Maclyn McCarty described a chemical process to extract from one strain of the *Pneumococcus* bacteria a long molecule—a "transforming principle"—that upon introduction into a different strain induced properties of the first strain, as identified by the presence of a particular molecule on the exterior surface of the recipient bacteria. The authors suggested that "the behavior of the purified substance is consistent with the concept that biological activity is a property of the highly polymerized nucleic acid"; they also commented, however, "In the present state of knowledge any interpretation of the mechanism involved in transformation must of necessity be purely theoretical."[13]

Avery et al. concluded the paper with the simple statement that "the evidence presented supports the belief that a nucleic acid of the [deoxyribose] type is the fundamental unit of the transforming principle of *Pneumococcus* Type III." Generalizing that assertion, and understanding the mechanism by which DNA encodes genetic information and can be used to transfer that information between organisms, constitutes a project that is, of course, ongoing. The natural language model of DNA as a "transforming principle" laid the foundation for every other model of the system, whether physical, metaphorical, algebraic, or graphical.

The ubiquitous modern cartoon representation of DNA is a ladder twisted around its long axis into a double helix that makes one revolution about every ten rungs. The rungs of that ladder are molecules that constitute genetic information. These molecules are often abbreviated by the first letters of their names, as A, T, G, and C. There are two complementary base pairs in each rung: an A paired with a T, a G with a C. A human genome consists of approximately 3 billion of these base pairs, separated into twenty-three pairs of chromosomes packed into the nucleus. Each chromosome is a contiguous molecule of DNA, many millions of bases long, in which the bases are organized into genes. Genes usually consist of about a thousand bases. Many genomes, human and otherwise, contain long stretches of DNA that do not appear to contain any genes whatsoever.

Information output from the genome begins with gene expression as follows: a polymerase molecule is a protein that reads a gene, proceeding in a particular direction along one of the DNA strands, transcribing DNA into RNA. One major function of RNA is to transfer information from genes in the nucleus to the cell's machinery for making protein. A gene is delineated from the rest of the letters in a chromosome by short sequences of bases at either end that tell the polymerase where to start transcribing and where to stop. This is, effectively, the molecular definition of a gene—it is the letters between the "start" and "stop" signals that are read by the polymerase, transcribed into RNA, and eventually translated into proteins. At this stage of understanding most genomes, this is in fact the only criterion for the identification of many genes. Little more is known about most human genes because it is far easier to read a gene than to determine its function. Indeed, we have not yet associated specific functions with most of the human DNA sequences identified via the molecular definition of "gene."

The word "gene" has another meaning, derived from classical studies of heredity, in which the notion of a gene is used to keep track of traits traced through generations. Classical genetic studies of hair and eye color or of diseases such as breast cancer and sickle-cell anemia demonstrate that for many traits there is a basic unit of heredity passed from one generation to the next. In their 1944 paper, Avery and his coauthors pulled together their experimental results with theoretical speculations of the day, noting that "the inducing substance has been likened to a gene, and the [molecule] which is produced in response to it has been regarded as a gene product." That is, the researchers were asserting that the "inducing substance" constituted an instruction for a new behavior, in that case, an instruction to make a particular molecule.

A major triumph of modern biology is the identification of some heritable traits with the information content of a sequence of bases, that is, with a single gene. Diseases such as sickle-cell anemia, cystic fibrosis, and certain breast cancers are each attributable to changes in a single gene. Diseases linked with a single gene are the easiest to identify; they are the low-hanging fruit of molecular medicine. Many traits, however, seem to be determined by several genes or the interaction of several genes. It will be many years before these complexities are understood. Until then, most researchers have their hands full sorting out what single genes do when expressed.

Returning to the cartoon description of gene expression, a polymerase transcribes DNA into RNA, which is in turn translated into protein by other molecular machines. Regulatory sequences of DNA adjacent to genes can influence how often, and under what conditions, they are expressed. Part of the "Central Dogma" of molecular biology, as Francis Crick called it in 1957, is that information flows from DNA to RNA to protein.[14] A working knowledge of the Central Dogma is enough to understand the elementary details of most current developments in synthetic biology. As this book progresses, I will point out examples in which the Central Dogma is insufficient to describe the workings of biology, and in which the cartoon must acquire another dimension in order to faithfully represent what we know about the living world.

The rest of this chapter utilizes only the simplest cartoon description of basic molecular biology. The description is built around three examples of designed biological systems that demonstrate various aspects of following four challenges: (1) defining the device physics of molecular components, (2) building and simulating models, (3) using defined components, and, most important, (4) building new biological entities based on quantitatively predictive designs.

Some synthetic biological circuits are more sensitive to uncertainty than others. In the first example below, the circuit operated as predicted by simulations in large part because the design was insensitive to unknown device physics. Constructed with a small number of unengineered moving parts, the circuit was designed to maintain a stable response to a signal after the signal ended. It is perhaps the simplest example of a biological system built to respond to its environment with a stable change.

The second example is a circuit that was less predictable because it was designed assuming a more-complete understanding of certain molecules than actually existed at the time. The circuit was supposed to exhibit predictable, periodic oscillation. This behavior required a more-complex simulation than the first example, and the simulation revealed that, to operate as intended, the design required modified proteins.

The third and final example involves building a design tool for a large number of parts in two complex organisms. It is an attempt to model viral infection at the level of single cells, an audacious first step toward building something akin to a flight model for a biological system.

Example 1: A Genetic Switch

In January 2000 Tim Gardner, Charles Cantor, and Jim Collins, all at Boston University, described in the journal *Nature* a genetic switch they built in the bacterium *E. coli*.[15] The switch is "bistable"—it has two stable states, on and off, controlled by a change in temperature or application of a chemical signal, in this case a molecule called IPTG. When the switch is on, the bacterium makes green fluorescent protein (GFP, taken originally from jellyfish) and glows green under ultraviolet light; when the switch is off, the protein is no longer made and any remaining protein slowly decays until the bacterium is dark.

Gardner and colleagues referred to this functional genetic element as a "genetic applet," named in analogy to small programs written in the computer language Java and designed to be self-contained and easily ported between computing platforms and operating systems. A genetic circuit element that behaves like a switch is likely to be a useful basic component in building or controlling more-complex biological systems.

The device physics of naturally occurring components was used to build a model.   Gardner et al. distinguished their approach from previous efforts by controlling the behavior of the genetic circuit through its *architecture* rather than by engineering specific parts to work in specific ways. They used a small network of naturally occurring proteins and genes and, with a quantitative model, predicted how those parts would interact to create a new desired behavior.

The molecular bits and pieces used in the switch were drawn from several different organisms. These parts had been used by many other people, in a variety of applications, but never before assembled to this end. Most important, the quantitative behavior of all of the individual parts was either previously characterized in *E. coli* or was measured by Gardner during his research, recapitulating the same progression—from natural language model to empirical model to quantitatively predictive model—that allowed the development of aviation. That is, the device physics of the proteins and genes was already adequately worked out. The individual description of each piece facilitated the construction of a quantitative model for the behavior of the pieces when combined in a circuit. Thus the circuit could be built both on paper and within an organism.

Circuit elements developed through evolution can be tricky to work with, but they are nonetheless understandable. The details of this work are worth considering. Not only is the science clever, but the construction of the switch lends insight into the style of genetic circuit elements that nature has developed over several billion years. Biology has settled on mechanisms that look convoluted at first glance. Some years down the road, it may be possible to construct molecules and mechanisms de novo that do exactly what we want in a straightforward way. Until then, we must make use of what we discover in existing organisms.

The fundamental genetic units employed by the researchers at Boston University are "inducible promoters," so named because as early as 1900 it was discovered that a signal from outside a cell can induce the cell to begin producing particular proteins.[16] The inducer does not work directly to enhance protein production. Rather, it works by interfering with a "repressor," a protein that inhibits transcription by binding to a region of DNA called a "promoter." A promoter sits just ahead of a gene, upstream of the "start" sequence, and is the location where a polymerase binds to DNA and begins transcription. By binding to a promoter, a repressor protein physically blocks access to the promoter so that the polymerase cannot read the gene. When the appropriate inducer signal is applied to the cell, it prevents the repressor from binding to DNA. Rather than directly acting to produce mRNA and protein, the inducer instead works through a double inhibition: a double negative, in other words. While we might wish for a simpler construction, with one moving part, evolution has settled on the more-complicated workings of inducible promoters. This requires more effort in both understanding and design. But these are the tools nature has provided.

By using a particular matched pair of repressor elements familiar to biologists for many years, Gardner, Cantor, and Collins were able to create a circuit with two stable states, in which either one or the other repressor gene was "on," producing mRNA and protein. This was implemented by creating a small, symmetric control loop on a circular piece of DNA. The team arranged the protein product of the first repressor gene (A) so that it inhibited transcription of the gene (B) coding for the second repressor. Similarly, the second repressor inhibited transcription of the first; the protein product of gene A inhibited gene B, and the protein product of gene B inhibited gene A. Flipping the switch from one state to the other was accomplished through the inducer signal, which interfered with the repressor protein that was currently "on," thereby allowing transcription of the other repressor.

Despite the obscuring nature of the biological nomenclature, the behavior of the circuit is simple and well defined. Heating up the cells turns the switch on, and adding the molecule IPTG turns the switch off.

The preceding paragraphs recount a *story* describing the behavior of the circuit. Correctly predicting *how much* inducer is required to flip the switch requires more effort, i.e., construction of a mathematical model based on the device physics of the repressor proteins and of the components of transcription and translation. Previous efforts to characterize the detailed device physics produced sufficiently accurate descriptions that the switch could be constructed with reasonable confidence that it would function as expected. As reported in *Nature,* the quantitative predictions of switch behavior based on the model were fairly accurate, but this is in part because the stable two-state switching behavior is not particularly complicated. The functioning of the switch was independent of some of the less well-known details of the device physics. Building a genetic circuit that displays behavior that continuously changes complicates matters considerably, as seen in the next example.

Example 2: A Genetic Oscillator

The second example was, remarkably, published in the same issue of *Nature* as the work on the genetic switch. Michael Elowitz and Stan Leibler, both then at Princeton University, built an oscillatory genetic circuit in *E. coli.*[17] Just as with the genetic switch, the raw materials came from a variety of organisms. Similar in basic design to the switch, the oscillator relied on inhibition of transcription, but with three rather than two inhibitory elements. Adding the third element created a longer control loop, and the inhibition proceeded around this loop with a distinct time delay between transcription events of any one of the three repressors; the protein product of gene A inhibited gene B, the protein product of gene B inhibited gene C, and the protein product of gene C inhibited gene A, in turn, around the loop. After the circuit was induced and began operating, it was intended to oscillate independently and indefinitely in each cell.

The reliance upon repression to make an oscillator prompted Elowitz to name the circuit the "repressilator." Transcription of GFP was again used to indicate the behavior of the circuit. Successful operation would be indicated by oscillating fluorescence in each bacterium. Perhaps because the repressilator circuit was designed for dynamic behavior, it was even more sensitive to imprecisely known aspects of the device physics, such as the binding affinities of repressor proteins for DNA and the lifetimes of repressor proteins within cells.

Simulation indicated that oscillation would require custom components:
Only rough design is possible with limited knowledge. Specific details of
the design were guided by simulation of a model, similar in structure to the
model used to develop the genetic switch at Boston University. The stable
protein production of the bistable switch made it insensitive to how long
those proteins lasted in the cell. Instead of producing stable behavior, the repressilator relies on constantly changing protein concentrations as part of its
internal dynamics. The repressilator simulation revealed that if the lifetimes
of the repressor proteins were too long, the circuit would produce stable expression of only one protein instead of sequential oscillations of all three.

As it turned out, the natural lifetimes of the proteins were in fact too
long. So Elowitz produced modified versions to ease the bridge between
modeling and experiment. The proteins were rebuilt with a small tag on
one end that enhanced degradation. The tag is recognized by protease enzymes in *E. coli*, which then destroy the attached protein, which is a con-
venient preexisting function to make use of in constructing a synthetic
biological circuit.

Not all simulations are created equal, because not all cells are created equal.
With new molecular components in hand—proteins tagged for relatively
rapid destruction by the cell—experiments proceeded. But differences between the experimental results and the simulation appeared immediately.
Both the amplitude and the oscillation period of the repressilator tended to
wander away from their initial values, with the peaks generally becoming
both higher and further apart. Here we see the blurring of model and simulation alluded to earlier in the chapter.

It turned out that the method used to simulate the model was an important aspect of accurate prediction. While the structure of the *model*—
differential equations in this case—allows for the initial numbers of molecules
to differ between cells, the original deterministic *simulation engine* allowed
no such variation; the concentrations of molecular species within a cell
were entirely determined before the simulation began. Because in reality
there may be only a few molecules of a particular kind in a cell, a small
fluctuation in the number of those molecules could potentially influence the
behavior of the circuit. This means the real system is subject to more noise
than would be allowed in a deterministic simulation. In order to produce
accurate predictions, the simulation method must reflect this variability.

The root problem is that even genetically identical cells do not produce
exactly the same number of any given protein, even under identical conditions. When there are 1,000 copies of a protein in one cell but only 990 in
another, it is unlikely the difference of 10 copies between the cells will cause

different behaviors. But if one cell has 15 copies and another only 5, this same absolute difference can have a significant biochemical effect within the cell. These differences in numbers can arise from a variety of causes: proteins can spontaneously decay, and there can be variation in the number of proteins made from a given copy of mRNA. Because the function of the repressilator depends on changes in protein concentrations, it is subject to this same problem.

In other words, the dynamic nature of the repressilator, in contrast to the stable expression of the switch, meant that the actual behavior of the genetic circuit could be significantly more sensitive to variations inherent in transcription, translation, and binding of the repressor proteins to the DNA. As described above, the original deterministic simulation method ignored potential fluctuations and indicated that the three repressor proteins would stably oscillate with equal amplitude (the same number of molecules) and equal period (the same amount of time).

A simulation method closer to physical reality would allow the number of molecules present at any given time to be subject to random fluctuations. This is more representative of the true physical situation inside a cell, or any small chemical reactor, than assuming a predetermined number of reactants. Elowitz built just such a simulation.

The improved simulation displayed behavior closer, though not identical, to the experiment. This means there are still features of the cellular machinery not yet included in the device physics. But the repressilator did display distinct oscillations in fluorescence, and that alone is sufficient to expect that future efforts will produce experimental results closer to predictions. It is difficult to say exactly why the repressilator's behavior does not mirror that of the simulation. There are still significant gaps in our knowledge of how most molecular components function. The natural language description of the oscillator is only slightly more complex than that of the genetic switch, yet the difference in the behavior of the mathematical model and the real system demonstrates that our ability to predict the behavior of synthetic biological systems is still limited.

Example 3: Rewriting a Genetic Program

The third example of fledgling biological design involves a virus, studied originally in the earliest days of molecular biology, that preys on *E. coli*. The long history of work on the virus, bacteriophage T7, is fertile ground for quantitative modeling. As described earlier, we still have a long way to go in defining even natural language device physics for the majority of genes and proteins in yeast, which have been studied for a longer period than T7. However, T7 is much simpler, and most of its components were described during the twentieth century.

While in graduate school at Dartmouth, Drew Endy began compiling all this knowledge about the various genes and proteins in T7 and the roles they play as the phage infects and destroys its host.[18] The effort began with a stack of research papers approximately two meters high, not including the unrecorded knowledge passed verbally from one researcher to another. His intent was to condense the many stories and individual measurements into a quantitatively predictive, comprehensive model of the infection cycle of the organism and its fifty-six genes.

Viewed in an electron microscope, T7 resembles a microscopic lunar lander. Infection begins when the landing legs on the virus recognize proteins specific to the surface of *E. coli,* and the phage "docks" to the bacteria's surface. The phage creates a pore in the cell wall of the host and the phage's genome enters into the host cell, whereupon the viral genes are transcribed sequentially in the order of entry. The infection process bears more than a passing similarity to standard computer programs, in which instructions are executed in a prescribed order.

As infection proceeds, resources of the bacterial host are consumed, with various structures dismantled for their valuable raw materials, and progeny phage are assembled from these components. Endy's model was designed to compute how many progeny result from a single infection event. But with only one data point for comparison—the original configuration of genes in the phage—any successful prediction made using the model could be a result of coincidence. So Endy embarked on a far more challenging test of the model.

After constructing and simulating a model of the native phage and comparing initial predictions from his model with results from the laboratory, Endy's next step was to find out what happens when the order of the instructions in the T7 genome were changed. In the model, this can be accomplished by changing around a few lines of code. An experimental testing involves laboriously constructing T7 genomes with genes removed from their original location and inserted elsewhere. This is genomic reengineering.

Endy and his collaborators first attempted this work in 1996, when synthesizing long sections of DNA was still both extremely expensive and technically challenging. The only methods available to Endy were "old-fashioned" techniques, in which genes are inserted at locations in the genome susceptible to enzymes that cut DNA at specific locations. New genes can be spliced into these gaps.

Several T7 genes transcribed early in the sequence work to shut down the host's own gene expression mechanisms so that the phage can hijack the host's cellular resources, providing the phage with critical raw materials. It thus makes sense that these genes are located near the head of the line. The position of other genes, however, seems less important upon initial inspection. And despite many decades of experience with T7, scientists have not completely described all of its genes. Based on previous genetic evidence, they fall roughly into the categories of "essential" and "nonessential" in their importance to the life cycle of the phage. Moving genes in the latter category, or interrupting their function via the insertion of other genes, should not affect the genetic program. Nonetheless, Endy found that interrupting several nonessential genes led to phages that grew poorly. Endy frequently notes that "nonessential" really just means "unknown."

Despite this inconsistency and complexity, Endy used the model to accurately predict the behavior of half of the experimental T7 strains (the wild type plus three artificial strains). Moreover, the model is compelling in that it does not contain any free parameters—no fudge factors that would facilitate matching experimental data with the simulation. The model can also be used to explore how the performance of the native configuration of genes compares with reordered configurations. After Endy examined the behavior of about one hundred thousand alternate genomes via simulation, it turned out that only about 3 percent seem to work better than the native configuration. Thus the order of the genes in the native phage seems well tuned, and in this respect nature seems to have done a good job of optimizing T7 for the environmental conditions in which it is found.

The T7 model is, for the moment, a unique tool to explore a complicated genetic system. Because it is the first effort of its kind, the model is not perfect. But improvement will come by finding out where it makes mistakes and why. To explore these mistakes, Endy is building a version of the T7 genome that is much easier to reorganize. Rather than relying on standard gene-splicing techniques, he is now synthesizing genomes from scratch—a technology that will be covered extensively in later chapters—in order to quickly test predictions from the model.

Endy's efforts were published in the *Proceedings of the National Academy of Sciences* in 2000.[19] He spent the next two years building better simulation engines and waiting for DNA-synthesis capacity to catch up to experimental requirements. Now at Stanford University, Endy and his collaborators have continued to build and test better models of phage T7.

The approach Endy has chosen requires fully characterizing those parts of T7 previously deemed nonessential and rebuilding the genome in a way that simplifies reorganization. He is now synthesizing a version of the phage that has short DNA "handles"—sequences that can be easily manip-

ulated using other molecular tools—at the end of many genes that facilitate moving those genes around in the lab. This version of the phage, T7.1,
also features a rewrite of the genetic code in which genes that overlap the
same section of DNA in the wild-type phage are, as Endy says, "un-
stuffed." This is an analogy to the software utility Unstuffit, which un-
packs compressed files. While the initial experimental demonstration of
T7.1 shows that a viable phage can be created this way, it remains unclear
whether the new sequence is easier to model or modify.[20] The success of
such dramatic reengineering is not guaranteed. Biology is by no means
simple, and leaping into a remodeling effort with some sixty different
genes is daunting. But the work of building complex genetic systems and
mathematical tools useful for understanding them is proceeding in earnest.

Conclusion

Whereas the flight model of a 747 contains descriptions of *all* the parts in
the aircraft, the three biological models I have described above consist of
only a handful of components. The device physics of each component was
determined primarily through extensive measurement rather than extensive theoretical knowledge based on first principles. The hope is that these
measurements take into account both (1) the unknown device physics of
all the other molecules present in a cell and (2) the influence of those molecules on the specific components described by the model. Of the three
models I used as examples, the T7 model is the most complicated, but it
still leaves out the vast majority of the constituent molecules in the host
cell. The switch and oscillator are more abbreviated still. It is as if the 747
flight model contained descriptions only of the control surfaces, without including details about structural components of the wing and the airframe.

Still, given the complexity of the task of constructing predictive models
of real biological systems, these efforts are an impressive start. The three
systems described above were developed within a few years of each other.
The scientists involved occasionally bumped into each other at meetings,
told each other stories about progress in the lab, read each other's papers,
and participated in the larger scientific community. In each of the examples,
experiments done on the lab bench were attempted only after constructing a quantitative model that provided predictions about the outcome. The
models therefore contributed to the choice of how the synthetic systems
were created in the laboratory. This is standard procedure for most other
fields of engineering but is entirely new in molecular biology.

This point is sufficiently important that it is worth repeating: the switch, repressilator, and T7 circuits were designed for particular functions with the use of rudimentary quantitative models. This is a trend in biology that will become more pronounced. New genetic circuits built from small numbers of moving parts will soon join the toolbox first occupied by the switch and the repressilator. The work with T7 takes the trend a step further, by trying to keep track of many more components whose behavior as a system is tightly constrained by decades of observations and millennia of evolution. Models that can be used to predict the effects of perturbations to existing biological systems will become de facto design tools for those components, providing an infrastructure to create new technologies based on biology. Future efforts will produce better results, as the molecular details governing such circuits are revealed.

The role of simulation in the three examples points to the future role of computers in biological engineering. Without computers, simulation of the genetic switch would have been hampered but possible nonetheless. Simulation of the repressilator without a computer would have been exceedingly difficult and for the T7 model would have been virtually impossible. As models of synthetic biological systems become more complex, computers will be crucial in exploring their behavior. But without validation through experiment, even Endy realizes his work on T7 would not be taken seriously.

Endy often notes he embarked on the project in part because, coming from the engineering community rather than the biology community, he did not know any better. In some sense, only an engineer would have attempted a modeling effort of this complexity. On the one hand, T7 has so many imprecisely described moving pieces that a physicist would have run away after one look under the hood, and the simulation involves so much mathematics that few, if any, biologists had sufficient training to even consider it. On the other hand, physicists Leibler and Elowitz had just the right background to conceive of the repressilator and to create a mathematical description of it.

Participation from biologists was, however, absolutely essential to the experimental tests of all three models. Little progress could have been made by the physicists and engineers without the accumulated store of knowledge of methods and biological lore that is part of the education and professional experience of molecular biologists. Such collaborative work is a sign of another clear trend: multidisciplinary approaches to biological problems are all the rage now. These efforts are transforming the way biology is studied, and will have similar impacts on biological engineering.

CHAPTER FIVE

A Future History of
Biological Engineering

Welcome, contestants, to another season of BioBrick Challenge! Your task today is to construct in ten hours a new microorganism that best completes the
challenge we will reveal in a moment. You'll be using DNA from the BioBrick
parts list, whatever you can find lying around the lab—don't forget to swab
the bench top and under the desks; people find the darnedest things in here—
plus whatever genes you can synthesize and assemble in the available time.
And remember, it isn't just the two teams here in our lab taking this challenge;
viewers at home with adequate facilities are competing too. Just e-mail us
your sequence, metabolic, and fitness data, along with video and photo-
graphs, and you have a chance to win fame and fortune. Well, fame anyway.
And I know we have some university and corporate labs in the challenge, as
usual, so good luck to you! But in the last ten years garage biology hackers
have scooped you every time.

THE BIOBRICK CHALLENGE is by no means a fantasy; it presently exists in the form of the International Genetically Engineered Machines
(iGEM) competition, coordinated by MIT. The iGEM competition is now
run every summer, with participating university students drawn from all
over the world. In 2006 several hundred students, organized into thirty-
nine teams from thirty-seven universities, represented countries around the
world.[1] In 2007 nearly seven hundred students from twenty countries com-
peted on fifty-nine teams.

The time scale for building a new genetic circuit during iGEM is not a few hours but rather several months. Yet this relatively short period of time is already evidence of a tremendous improvement in technology over just the last five years. IGEM originated in a course run at MIT during the Independent Activities Period (IAP) between academic terms. During the first go-round, in 2003, getting the designs assembled and tested required most of a year; none worked as predicted. Participants in the second IAP course finished off the job in about six months, yet again produced no functional circuits. Now teams in the iGEM competition begin their projects in early summer, after classes are finished, with a deadline of early November for the iGEM Jamboree, when all teams show off what they have accomplished.

The reason the iGEM competition runs over months instead of hours is that the competitors are presently manipulating ill-defined parts with tools poorly suited for the job. This is the same reason that building the biological circuits examined in the last chapter required so much effort. The current generation of biological tools is completely unsophisticated, even compared with the sets of screwdrivers, wrenches, and finely divided Allen keys found in most home garage toolboxes. Similarly, the present list of biological parts is completely unrefined in comparison with the standardized screws, bolts, and nuts found right next to the toolbox. Yet the improvement from the person-*years* required to produce the repressilator and the genetic switch, described in Chapter 4, to the person-*months* for the iGEM teams is a groundbreaking achievement. An increasing number of student projects from iGEM have either themselves been published in academic journals or have produced parts that enabled experiments that were published.

The main reason for the substantial improvement in productivity is the existence of the Registry of Standard Biological Parts. The Registry is a list of biological parts and the conditions under which they have been characterized. The parts exist in the Registry, and are maintained as stocks in a freezer, as genetic sequences with ends that can be pasted together using defined protocols both on the computer and in the test tube. One goal of the Registry is to compile a list of useful circuit elements, in which the entry for each element contains information about where it has been used and design guidelines for incorporating it into new designs. The Registry and iGEM, as imagined by the organizers, are part of a grand experiment:

> The iGEM competition and the Registry of Standard Biological Parts (aka "the Registry") are practically testing the idea that biological engineering can be made reliable through the use of freely-shared, standardized, and well-documented parts called BioBricks. [BioBrick parts] have special features that allow for iterative assembly of longer "composite parts," leading to larger devices or systems.

iGEM teams are challenged to create working devices by designing and assembling a) parts that already exist in the Registry and b) their own parts that conform to the BioBrick standard. New parts must be documented in the Registry database and physically sent to the registry. This allows future biological engineers to build on the creativity, expertise, and experience of the people and groups that came before them.[2]

Given the progress demonstrated in the competitions, the experiment appears to be succeeding.

The Registry of Standard Biological Parts and iGEM focus on the role of parts and their assembly as a contribution to developing the engineering of synthetic biological systems. This is partially because many parts exist in refrigerators and freezers around the world as genes on plasmids. These genes are, in effect, biological LEGOS ready to be picked up and played with. The ability to add novel BioBrick parts to the bin adds to both the fun of iGEM and the utility of the Registry. Making no bones about the analogy, the organizers of iGEM celebrate the grand-prize winners with a trophy in the form of a giant aluminum LEGO brick.

Yet for iGEM to become the BioBrick Challenge imagined at the opening of this chapter, considerable effort must also go into developing not just parts but the tools to characterize and then assemble those parts into functional circuits. This is no summer project.

The iGEM competition is run for university students, which provides another reason to concentrate on parts rather than new tools. Developing new tools can be extraordinarily difficult and time consuming, whereas students in iGEM need to get something working in just a few months. There presently exist sufficiently capable tools to manipulate and assemble BioBrick parts into interesting combinations. However, the capabilities of these novel systems will be limited without much more detailed and accurate descriptions of the device physics of the parts, new means of analyzing and modeling complex synthetic systems, and even more-capable tools to read and write genomes. This is the lesson to be learned from examining the histories of other technologies.

The Technology Line

The developmental paths of modern technologies display common trends. These trends can be generalized by comparing examples of technology development. The trends can then be divided into stages defined by the ability, or lack thereof, to quantitatively predict the behavior of human artifacts. When applied to a relatively new, or immature, technology, such as biology, this analysis can highlight "what needs doing" in order to develop infrastructure to produce economic and social benefits. As I describe below, a comparison of the development of multiple technologies suggests that investment in particular capabilities has predictable consequences.

Over the next few pages, I introduce a "Technology Line" as a new tool to understand the progression of technologies not as a function of time or place but rather as a function of capability—in particular, the capability to design, construct, and test functional objects that have material value and utility. In other words, a Technology Line helps contrast the present capabilities of a new technology with capabilities that require further investment. Thus a Technology Line may serve as a guide to the future of any given technology.

Underlying the technologies we take for granted are tools used to design, build, and test products. This infrastructure governs the transition from *ideas* with potential value to *things* in the world that work. Understanding the importance of tools provides a framework for considering broad questions about the evolution of biological technology. What, for example, needs to be built or invented to enable biological technology to look like aviation, a wildly successful example of modern technology development?

The following narrative was derived primarily from examining the histories of computers and airplanes. Aviation, as we know it today, developed within approximately a century and is an archetype for modern technology development.

Figure 5.1 is a representation of important personalities and events in the history of aviation. The time scale is not linear (i.e., not to scale). Although the timeline in figure 5.1 provides a way to judge the time required for different stages of development for aviation in particular, generalizing the progression is not possible without identifying important transitions between stages of experimentation, design, and construction. Removing the dates altogether, as in figure 5.2, allows comparison of different technologies by providing a means to think about invention and development more generally.

Replacing the timeline with the Technology Line is a way to visualize how technologies change. It is another nonlinear cartoon—itself a natural language model—that represents how stories about technology evolve in time. These stories reflect our current understanding of the world or our understanding of a thing we are building. The stories start out composed in natural language, that is, a narrative. Natural language is then displaced by empirical descriptions based on quantitative data and experimentation. These in turn are displaced by theories developed using contemporary mathematics and physics, finally become design tools based on predictive

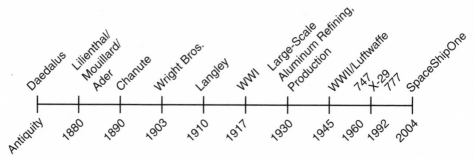

Figure 5.1 A Timeline (not to scale) of important personalities and events in the history of aviation. See Chapter 4.

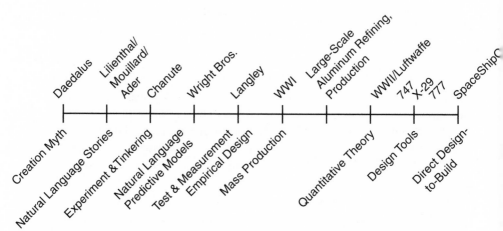

Figure 5.2 A Technology Line for aviation, with dates replaced by phases of technology development.

models, and are used in direct design-to-build processes. The last stage describes current design and manufacturing processes in the fields of aviation and computers, and a comparison reveals how far biological technologies have to go. The progression of stories is characterized at each stage by the appearance of tools that enable manipulation of the subject matter, which could be atoms or bits. The Technology Line is thus a metric for technology based on progress toward implementing a design-to-build infrastructure, which is, in some sense, the culmination of the project called engineering.

The progression of computer and communications technology is another illustrative example that can be mapped on a Technology Line (see figure 5.3). The timeline for critical events in the history of computers is

Figure 5.3 The Technology Line for computers is not a strict representation of the order of events; important transitions appear nonchronologically. Unlike aviation, quantitative theory and design emerged in computers before the appearance of mass production of integrated circuits. There is, however, a general progression in complexity and capability from left to right. Moreover, elements on the left are clearly prerequisites for elements on the right.

hundreds of years longer than that for the history of aviation. Yet the transition from the arrival of gear that enabled progress from empirical design and test and measurement to the design-to-build world took only around forty years in each case. There is no deep surprise here: many of the developments in computation were pushed by aerospace defense requirements during the cold war. Returning to the theme of biological technologies, we can see that just as the ability to simulate the physical behavior of complex systems is crucial to both computers and aviation, it is likely to be a fundamental requirement for successfully building complex biological systems.

The Technology Line is designed to build analogies between the evolutionary paths of human technologies, and to reveal differences between them. When looking back in time, this strategy imposes a retrospective structure on a process inevitably full of twists and turns. As Scott Berkun notes in *The Myths of Innovation*, "mistakes are everywhere, rendering a straight line of progress as a kind of invention itself."[3] Moreover, there are events along some lines that are not repeated on other lines or that appear in a different order. Nevertheless, there are clear similarities between the transitions in computer engineering and in aviation. Assembling Technology Lines illustrates the *kind* of technology that appears at transitions between types of stories. By elucidating the details of particularly important innovations, a compilation of Technology Lines thereby suggests what is yet to be introduced in the development of immature technologies such as biology.

The short summary, as evidenced by the first chapters if this book, is that the community of scientists and engineers working to build biological systems has a long road ahead. By comparing the Technology Line for biology with those for aviation and computing (figure 5.4), it appears biological technologies have strong records in tinkering and the beginnings of empirical design, whereas there is considerable further effort required to develop design tools and implement the direct design-to-build infrastructure. The current limited ability to control genetically modified systems suggests that biological technologies are closer in development to Otto Lilienthal's glider than to even the Wright Flyer.

To be clear, I make no claims about predicting the specifics of a technology's development nor the time scales over which development proceeds. But there are important design and manufacturing capabilities that characterize how technologies move from dreams to descriptive stories to objects that perform according to exacting specifications, at every step becoming more sophisticated and capable products that contribute to our society and economy.

Because the Technology Line is itself a natural language model describing the process of technology development, it is worthwhile to ask if there will ever be a quantitative and predictive model of technology development. As that would require the ability to predict the arrival of specific personalities, collisions of ideologies that result in conflict, and even new physics, I have doubts about development of any such future history. It is by no means clear that there *can be* a unified description of innovation or of the evolution of technology. Specific technologies arise in the context of history and the various social, economic, and political pressures of the day.

Today's environment is defined by rapid communication of ideas, rapid simulation of those ideas by using software, and finally the rapid fabrication of objects. Because we have experience with several hundred years of new technologies in the context of the industrial and information technology revolutions, we can at least say something empirical about how much time is required for a design-to-build infrastructure for any technology to arise, how much time tends to pass for a technology to be widely adopted, and how long it generally takes to realize subsequent economic value.

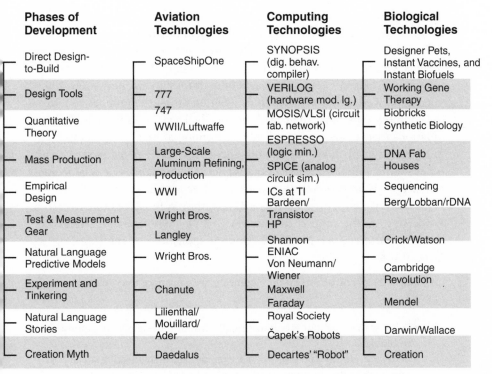

Phases of Development	Aviation Technologies	Computing Technologies	Biological Technologies
Direct Design-to-Build	SpaceShipOne	SYNOPSIS (dig. behav. compiler)	Designer Pets, Instant Vaccines, and Instant Biofuels
Design Tools	777	VERILOG (hardware mod. lg.)	Working Gene Therapy
Quantitative Theory	747 WWII/Luftwaffe	MOSIS/VLSI (circuit fab. network) ESPRESSO	Biobricks Synthetic Biology
Mass Production	Large-Scale Aluminum Refining, Production	(logic min.) SPICE (analog circuit sim.)	DNA Fab Houses
Empirical Design	WWI	ICs at TI	Sequencing Berg/Lobban/rDNA
Test & Measurement Gear	Wright Bros. Langley	Bardeen/ Transistor HP	
Natural Language Predictive Models	Wright Bros.	Shannon ENIAC	Crick/Watson
Experiment and Tinkering	Chanute	Von Neumann/ Wiener Maxwell	Cambridge Revolution
Natural Language Stories	Lilienthal/ Mouillard/ Ader	Faraday Royal Society	Mendel
Creation Myth	Daedalus	Čapek's Robots Decartes' "Robot"	Darwin/Wallace Creation

Figure 5.4 A Comparison of Technology Lines for aviation, computers, and biology.

The Structure of Modern Technological Revolutions

The culminating step in the Technology Line appears to be the achievement of a design-to-build infrastructure. Parts are designed in computer-aided design (CAD) using sophisticated engineering models based on real-world data, then produced in a fabrication shop, often using computer-aided manufacturing (CAM) machinery, and then shipped to the designer in days or hours. Design-to-build is not part of a general story about technological evolution but rather a story restricted to primarily post-nineteenth-century invention. Something fundamental changed during the twentieth century: the ability to simulate the performance of an object became powerful enough to predict the behavior of the object before it was built. Bits and atoms became fungible, in both physical and economic senses of the word; namely, computational descriptions of things can be interchanged with the things themselves for the purposes of design and testing, while increasingly those designs are sold with the understanding that they can be instantiated by anybody with the right computer-aided manufacturing tools.

The ability to abstract and simulate also facilitates exploration of new designs. Of greater importance, the proliferation of composable parts provides a consequent capability to combine those parts in new ways. Once you can join elements with different functionalities, the number of potential combinations starts to grow dramatically. Devices resulting from combinations themselves become elements, forming a (recursive) hierarchy of utility. It is this profusion of functionality that is likely to lead to the greatest economic impacts. But just because new functionality is available does not mean it is immediately put to use in the economy.

James Newcomb, my colleague and coauthor at Bio Economic Research Associates, set the context this way in *Genome Synthesis and Design Futures: Implications for the U.S. Economy,* sponsored by the U.S. Department of Energy:

> Taking a long view of current developments, we can address several key questions:
>
> • What is the potential economic significance of the engineering approach to biology?
> • What are the attributes of the emerging biological engineering revolution and how do these compare with previous technology revolutions?
>
> Almost everywhere we look, technology exhibits patterns of recursiveness: technologies are constructed from parts that already exist. These parts, in turn, are comprised of similar combinations of sub-parts—a pattern that repeats itself across as many as five or six layers. W. Brian Arthur, an economist and technology thinker at the Santa Fe Institute, observes that "technology forms a set—a network—of elements from which novel elements can be constructed. Over time, this set builds out by bootstrapping itself from simple elements to more complicated ones and from a few building-block elements to many."
>
> By enabling innovation through combination across a wide range of biological components and modules, synthetic biology could radically change the landscape of biotech innovation. The power of innovation through combination of existing components is already demonstrable in technology domains such as combinatorial chemistry, electronics, and software, where decades of innovation have built on previous developments. ... Systems such as these that foster prolific innovation through combination are not just technical; other preconditions include economic, social, and regulatory frameworks that determine the appropriability of value by innovators and intellectual property systems that can support the creative accumulation of innovations over time.[4]

Technology revolutions are not just about tools, methods, and the ability to produce an object with certain functions. To realize value, whether from

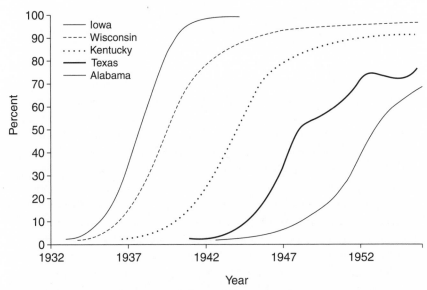

Figure 5.5 Adoption rates of hybrid corn across the United States.

an innovation in health care, food production, or materials fabrication, or even from a simple toy, people must be ready and able to understand, acquire, and put to use those innovations.

An example of technological adoption with profound impacts on food production and wealth creation in the early to mid twentieth century was the gradual adoption of hybrid corn by farmers in the United States. Hybrids are developed through breeding rather than direct genetic modification. Although farmers working in different states had access to roughly the same information and technology from the 1930s onward, hybrid corn varieties were developed and utilized at quite different rates, depending on multiple factors (figure 5.5).

Michael Darby and Lynne Zucker, who study innovation at the University of California at Los Angeles, generalized the example of hybrid corn to describe technology revolutions of different kinds. They write extensively about nanotechnology and biotechnology and highlight the success of corn as a particular kind of innovation:

> Zvi Griliches was the first economist to study the class of breakthrough discoveries which he named an "invention of a method of inventing." His case study was hybrid corn, "a method of breeding superior corn for specific localities. It was not a single invention immediately adaptable everywhere." Griliches observed that such breakthroughs thus involve a double-diffusion process: the timing of application of the inventing method to specific potential products

("availability") and the speed with which sales of each specific product reach a mature level ("acceptance"). Griliches related the speed of both processes to their profitability.[5]

Griliches and, later, Darby and Zucker have placed significant weight on the profitability of a method of invention in the case of corn. Also crucial were the identification of desirable traits, the ability to predictably manipulate them, and the willingness by farmers to exchange currency and control of knowledge for improved yields.[6] All of these factors contributed to the several decades it took for hybrid corn to reach high levels of use.

Diffusion lags in acceptance have appeared frequently in the adoption of new technologies over the last several centuries. After the demonstration of electric motors, it took nearly two decades for penetration in U.S. manufacturing to reach 5 percent and another two decades to reach 50 percent. The time scale for market penetration is often decades (see figure 5.6).[7] There is, however, anecdotal evidence that adoption of technologies may be speeding up, as Newcomb, Carlson, and Aldrich note: "Prices for electricity and motor vehicles fell ten-fold over approximately seven decades following their introduction. Prices for computers have fallen nearly twice as rapidly, declining ten-fold over 35 years."[8]

Regardless of the time scale, technologies that offer fundamentally new ways of providing services or goods tend to crop up within contexts set by preceding revolutions. The interactions between the new and old can create unexpected dynamics, a topic I will return to in the final chapter. More directly relevant here is that looking at any given technology may not give sufficient clues as to the likely rate of market penetration. For example, while the VCR was invented in 1952, adoption remained minimal for several decades. Then, in the late 1970s, the percentage of ownership soared. The key underlying change was not that consumers suddenly decided to spend more time in front of the television but rather that a key component of VCRs, integrated circuits, themselves only a few decades old at the time, started falling spectacularly in price. That same price dynamic has helped push the role of integrated circuits into the background of our perception, and the technology now serves as a foundation for other independent technologies, ranging from mobile phones to computers to media devices.

Similarly, the overlap of two technologies new to biological engineering is completely transforming the way synthetic genetic circuits are built and eventually how new organisms will be built. The creation of a library of composable genetic parts, with mathematically abstracted functions that can be modeled in software, and the ability to turn electronic representations of circuits into molecules via chemical DNA synthesis are together

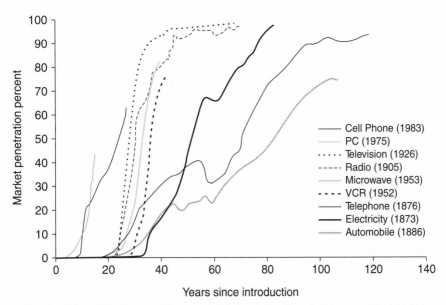

Figure 5.6 The diffusion of technologies can take decades (as measured by market penetration). The data set extends only to 1997. Source: P. Brimlow, "The silent boom," *Forbes,* 7 July 1997.

beginning to enable de novo construction of pathways to produce within a matter of weeks materials with both economic and social value. This methodology is an example of a context set by a preceding revolution, that of information technology and communications, a context which provides the capability to model the behavior of new circuits and to send out electronic specifications that are shortly returned via express mail. The adoption of well-characterized parts with defined functions dramatically alters the way engineers think about building synthetic systems, and the burgeoning ability to exchange matter (atoms) and information (bits), within a few days, will make the whole project of biological engineering work within the context of the economy.

The Fungibility of Physical Information

Bits and atoms are so fungible in some areas of design and manufacturing that three-dimensional scanners are used to turn objects into digital information. Some automobile designers prefer to sculpt a new svelte shape by hand in clay rather than start the process on a computer. The size and shape of the lump of clay is then precisely measured by an automated system that generates an electronic design file: from atoms to bits.

People working with biological parts have also generally exploited the atoms-to-bits direction. Literally taking pictures of molecules and processes is now the norm, with that data then described by models of varying sophistication, to be debated and hopefully understood. It is only recently that biological technologies have enabled a facile implementation of bits-to-atoms, in which electronic information is turned into physical pieces of DNA that code for new genes and organisms.

The relevant biological technologies are emerging and improving so quickly that it is difficult to grasp how our capabilities are changing. The Technology Line in figure 5.4 (page 57) reveals opportunities for investment in the fundamental infrastructure of biological engineering that could influence virtually the entire scope of human economy and industry. Moreover, the Technology Line places progress in biology within a broad technological context that can be supplemented by examining specific changes in cost and productivity. Estimating the pace of improvement of representative technologies is one way to illustrate the rate at which our ability to interact with and manipulate biological systems is changing. As I will explore in the next chapter of this book, the pace of improvement underlies the explosive potential for promise and peril as we "learn to fly."

CHAPTER SIX

The Pace of Change in Biological Technologies

ALL I WANT FOR CHRISTMAS is my Discovery DNA Explorer Kit![1] With a colorful plastic centrifuge, a few bottles of simple chemicals, and a price tag of $79.95, anyone can get started manipulating DNA from any organism at hand. A further quick trip online will suffice to procure enzymes and reagents for cutting, pasting, and amplifying nucleic acids. Children can now begin playing with DNA in their bedrooms, and no doubt hacking will soon follow. Those truly motivated to hack DNA could do so now by ordering the kits often used in academic or industrial labs and by compiling information from the Internet. There is no mystery in extracting DNA from one organism and inserting it into another. Success simply requires a willingness to keep trying until the procedure works—though, as we have seen, getting DNA to behave as desired in its new environment is still somewhat problematic. Nonetheless, the fundamental operations required to transfer DNA from one organism to another (sometimes referred to as "bashing DNA," a term providing some insight into how the technology has been viewed within the laboratory setting) are now well and truly a part of common knowledge. Many basic laboratory routines of molecular biology have been reduced to recipes that anyone can follow.

The lesson is that the advent of the home molecular biology laboratory is not far off. The physical infrastructure for doing molecular biology is becoming more sophisticated and less expensive every day. Automated, commercially available instrumentation today handles an increasing fraction of laboratory tasks that were once the sole domain of doctoral-level researchers, thereby reducing labor costs and increasing productivity. This technology is gradually moving into the broader marketplace as laboratories upgrade to new equipment. Used, still very powerful instruments are finding their way into wide distribution, as any cursory tour of eBay or other online clearinghouses will reveal. Anyone can easily outfit a very functional lab for less than $5,000 with equipment that just a few years ago cost at least ten times as much.

The proliferation of protocols and instrumentation will soon put highly capable tools in the hands of professionals and amateurs worldwide. More important, the general improvement of technologies used in measuring and manipulating molecules will soon enable a broad and distributed increase in the ability to alter biological systems. The resulting potential for mischief or mistake is causing understandable concern—there are already public calls by scientists and politicians to restrict access to certain technologies, to regulate the direction of biological research, and to censor publication of some new techniques and data. It is questionable, however, whether such efforts will in fact increase security or benefit the public good. Proscription of information and artifacts generally leads directly to a black market that is difficult to monitor and therefore difficult to police.

I will argue through Chapters 9–14 that an alternative superior to restriction is the deliberate creation of an open and expansive research community. That open community may be better able to respond to crises and better able to track research, whether in the university or in the garage. First, we have to understand the state of the technology and the capabilities it gives those who wield it.

Factors Driving the Biotech Revolution

New technologies are announced every day that enable better access to the molecular world. These tools do much more than simply improve the abilities of scientists to gather data.

Beyond its own experimental utility, every new measurement technique creates a new mode of interaction with biological systems. Moreover, new measurement techniques can swiftly become means to manipulate biological systems. Technology is emerging and improving so quickly that grasping how our capabilities are changing is difficult. Estimating the pace of improvement of representative technologies is one way to illustrate the rate at which our ability to manipulate biological systems is advancing.

It is not easy to quantitatively assess the rate of improvement of many technologies. The effort is aided when instrumentation companies advertise the prowess of their machines. In addition, some research papers describe in detail how fast these instruments work and how much they cost to operate. Of particular interest, manipulating DNA has become so common that speed is critical in many applications. For example, chemically synthesized DNA fragments, or oligonucleotides, can be used in DNA computation, in the fabrication of gene expression arrays ("gene chips"), and in the creation of larger constructs for genetic manipulation. DNA for these purposes is synthesized either in the lab where it is to be used or elsewhere in a dedicated facility and delivered by express mail. Mail-order oligonucleotides were used in 2002 to build a functional poliovirus genome from constituent molecules for the first time.[2] Since then, constructs of ever-increasing length have been announced (figure 6.1), and a full bacterial chromosome was recently synthesized from short oligos.[3] Access to gene-length synthetic DNA is crucial to the iGEM competition, as discussed in Chapter 5.

Assembling synthetic genes and genomes would be technically implausible and financially prohibitive without large-scale oligonucleotide synthesis. Demand for synthetic DNA comes additionally from companies developing new detectors in antibioterror devices, new medical diagnostics, and new genetic circuits and new organisms for the production of materials and fuels. Thus the rate at which DNA-synthesis capacity is changing is a measure of the improvement in our ability to manipulate biological systems and biological information. Delivery times for synthetic constructs are now a rate-limiting step in product development at some companies. Measurements of productivity improvements in DNA synthesis are therefore direct measurements of economic value. Similarly, improvements in reading the code of existing pieces of DNA are a measure of our ability to gather biological information in diverse applications. Sequencing is now a crucial tool in understanding human health and disease, detecting new natural and synthetic threats, prospecting for new parts from nature, and proofreading synthetic circuits and organisms.

The Pace of Technological Change in the Context of Moore's Law

The historical development of integrated circuits provides an exceptionally useful analogy for what is happening in biology. Indeed, a cadre of engineers—many of whom began their professional lives within the computing revolution launched by the capability of fabricating high-density microelectronic circuits and then by the proliferation of design tools and batch fabrication—are now pushing to replicate this infrastructure in biology. However, there are additional profound similarities, and differences, between these fields that deserve exploration.

The simplest way to gauge progress in biological technologies is to plot

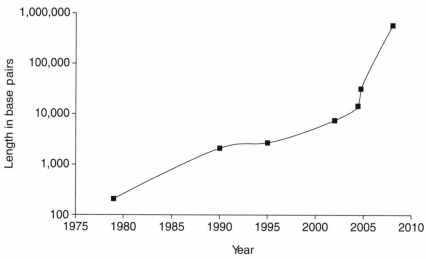

Figure 6.1 Longest-published synthetic DNA constructs.

changes in productivity enabled by commercial instrumentation. Because many of these instruments are automated, it is possible for one person to look after more than one at a time, thereby amplifying that individual's capabilities in the lab.

Figure 6.2 contains estimates of potential daily productivity of DNA synthesis and sequencing based on commercially available instruments, including the time necessary to prepare samples. There have been only a few generations of instruments, and these estimates are intended primarily to capture the essence of the trends. As a comparison, the number of transistors on microchips, is also shown in figure 6.2. Improvements in computing power have long been measured by tracking the number of transistors that can be fabricated on an individual chip. The exponential increase in transistor numbers visible in figure 6.2, known as Moore's Law, was first quantified by Gordon Moore, one of the founders of Intel, in 1965.[4] Improvements in DNA synthesis and sequencing appear to keeping pace with Moore's Law.

These trends are reflected in other areas of biological research. The process of determining protein structures, which relies on a variety of technologies and fields of expertise, from genetic modification to chemistry to X-ray crystallography to computer science, has experienced a similar dramatic improvement in productivity (see figure 6.3, which depicts time to completion rather than productivity). This suggests a broad and general rapid improvement of biological technologies.

Comparing any benchmark to Moore's Law has been a cliché for some years now, but doing so remains a useful device to gauge our expectations

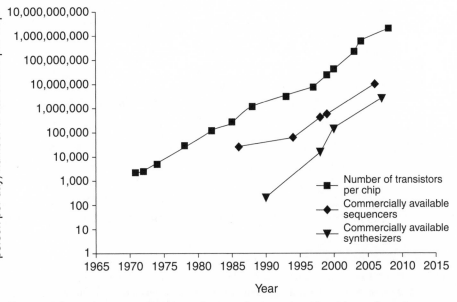

Figure 6.2 Productivity in DNA synthesis and sequencing. On this semilog plot, DNA-synthesis and DNA-sequencing productivities as enabled by commercially available instruments are rising approximately as fast as the number of transistors per chip, a frequent surrogate measure of productivity in the wider economy. Productivity is defined here as the amount of DNA that can be processed by one person running multiple machines for one eight-hour day, as set by the time required for preprocessing and sample handling on each instrument. Not included in these estimates is the time required for sequence analysis. This data was originally published in R. Carlson, *The pace and proliferation of biological technologies,* Biosecur Bioterror 1, no. 3 (2003): 203–214, and is supplemented by manufacturers' data for more-recent instruments.

of how technologies other than computers will affect socioeconomic change. Any differences that can be identified are particularly interesting. In this case, Moore's Law emerges because chip doubling times are a consequence of the planning intrinsic to the semiconductor and computer industry. The most succinct statement of the Law in Moore's original paper on the subject is that "The number of transistors per chip that yields the minimum cost per transistor has increased at a rate of roughly a factor of two per year."[5] The "minimum cost per transistor" is a function many different factors, including the cost of raw materials, the minimum size for a transistor created using any given generation of manufacturing technology, the failure rate for that generation of manufacturing technology, and the cost of packaging the raw chip in a way that can be used by computer manufacturers. That is, at its very origins, Moore's Law emerged from a statement largely about cost,

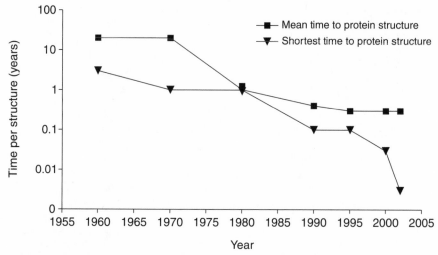

Figure 6.3 Estimated time to determine protein structure (isolation/production, crystallization, data collection, model building). The dramatic reduction in the time required to determine protein structures is evidence of a general trend of increased productivity in biological technologies. Many of the technologies used in finding protein structures are widely used in biology for other purposes. The shortest time and mean time to find protein structures were computed from a survey of five crystallography labs (R. Carlson, *The pace and proliferation of biological technologies,* Biosecur Bioterror 1, no. 3 (2003): 203–214). The time required for each step can vary significantly, depending upon the protein. For example, successful crystallization may take anywhere between hours and months of effort. The difference between the estimates of the average time to structure and the shortest time to structure illustrates the difficulty in absolutely quantifying productivity.

and more broadly about economics, rather than a statement strictly about the dictates of physics and chemistry, or about the limits of technology.

On a deeper level, Moore's Law is primarily a function of the capital cost and resource allocation necessary to build chip fabrication plants. It is true that progressively more sophisticated manufacturing technology was required to increase the density of transistors with each generation of chips, yet the time scale for doubling that density was not intrinsically set by technological progress but rather by the time it took to pay off the cost of building the fabrication plant. The payment period, in turn, has always been a function of projected income. Ultimately, therefore, the pace of Moore's Law is set by demand. Projected revenues enable Intel and its competitors to finance the enormously expensive infrastructure needed to build chips. Moore's Law is a description not of what is technologically possible, but of what is economically plausible. Therefore, Moore's Law, and any other measure of productivity, even if indirect, tells us about the capacity for real work by human hands in the economy. That real work can in turn enable even greater changes in productivity.

As each new generation of computer chips is capable of handling greater computational tasks, those capabilities enable the design and simulation of yet more-powerful subsequent chips. In other words, for most of the last thirty years there has been feedback between the manufacturing of one generation of chips and the design power they could bring to manufacturing the next generation of chips.

We can now see the beginnings of a similar effect in the development of biological technologies. For example, modified enzymes, optimized for specific tasks in the laboratory, are used to prepare DNA for sequencing and are themselves products of earlier sequencing projects. Recombinant proteins are used every day to elucidate interactions between proteins within organisms, and that information is already being used to design and build new protein networks. Enzymes are directly used in a sequencing process known as pyrosequencing.[6] This technology, when combined with microfluidic and optical instrumentation, was recently used to resequence the genome of James Watson, the codiscoverer of the structure of DNA, for approximately $1 million. (This effort is known as Project Jim, and is an improvement that keeps up the previous pace toward the thousand-dollar human genome. See figure 6.5 and the discussion below.)[7] The acceleration of capability demonstrated in Project Jim is an indication of what will happen when we begin to manipulate biology by using biology on a large scale at many levels of complexity.

Previous observers have compared increases in the total number of sequenced genes to Moore's Law. But this mixes proverbial apples and oranges, because total sequencing productivity is a measure of total industrial capacity (the number of sequencing instruments produced and in operation), whereas the number of transistors per chip is ostensibly a measure of the potential productivity enabled by each individual computer. The total number of sequenced genes is more analogous to the total number of computer chips in existence, or possibly the total number of computational operations enabled by those chips. Comparing Moore's Law to estimates of the daily productivity of one person at a biology laboratory bench is appropriate because that productivity determines how much benefit, or havoc, one person can generate.

An alternative statement of Moore's Law is "computational resources for a fixed price double every eighteen months." If we assume for a moment that the cost of appropriately skilled labor has remained constant, the units of the vertical axis in figure 6.2 (bases synthesized and sequenced *per person per day*) match the metric of resource cost, which is explicitly labor in this case. The capability of a single individual in the laboratory has improved dramatically over the last fifteen years. Note that the assumption of fixed labor cost is too conservative. Labor costs associated with sequencing have fallen as benchtop laboratory techniques that once required a doctorate's worth of experience have been replaced by automated processes that can be monitored by a technician with only limited training.

The Human Genome Project was originally forecast to require approximately fifteen years and $3 billion.[8] Because of advances in automation, biochemistry, and computation, the job was finished two years ahead of schedule and significantly under budget. Changes in technology and technique enabled the Celera Genomics, at the time a subsidiary of the Applera Corporation, to accelerate dramatically the completion of a rough draft of a human genome, albeit using a great deal of information provided by the public project. Technology development was part of the original Human Genome Project road map, and money was always available to buy many slow machines under the original plan for the publicly funded genome project. Coordinating the effort and paying for the labor to run those machines was initially prohibitively expensive for a private project. However, the advent of new technology provided for rapid increases in sequencing productivity, which allowed Celera to launch its private program. A commercial effort was feasible only when sequencing instruments and sample preparation became sufficiently automated, whereupon one person could shepherd several machines and the total task could be completed in an interesting time interval. This required highly centralized sequence-production facilities in order to minimize the number of instruments, given their individual high cost. This infrastructure is similar to that of microchip fabrication plants, otherwise known as "chip fabs."

However, sequencing instruments have become much closer to commodities than have the plasma etchers and vapor deposition systems used in microchip production. Sequencing machines are already widespread in laboratories, and there is clear demand for faster, cheaper instruments. There is every reason to expect new technologies to continue the increase in productivity and the decrease in cost. Consequently, it is not at all obvious that the current centralized-service model will be relevant to the future of biological technology. Therefore, it seems likely that biological discovery and the technology it enables will continue to result in lower-cost, highly capable instrumentation.

More significant, the long-term distribution and development of biological technology is likely to be largely unconstrained by economic considerations. Costs for making chips continue to rise; costs for reading genomes continue to fall. While Moore's Law is a forecast based on understandable large-capital costs and projected improvements in existing technologies, which to a great extent have determined its remarkably constant behavior, current progress in biology is exemplified by successive shifts to new tech-

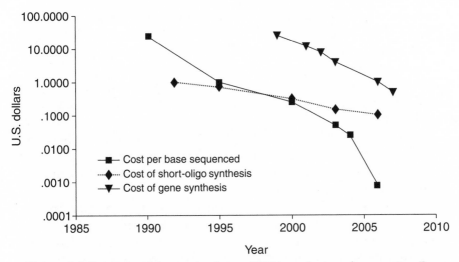

Figure 6.4 Estimates of the cost per base of DNA synthesis and sequencing. For data sources, see Carlson, *The pace and proliferation of biological technologies*, Biosecur Bioterror 1, no. 3 (2003): 203–214.

nologies. These technologies share the common scientific inheritance of molecular biology, but in general their implementations as tools emerge independently and have independent scientific and economic impacts.

The cost of chip fabs has now reached more than $6 billion per facility and is expected to increase, but there is good reason to expect that the cost of biological technologies will only decrease.[9] Indeed, the continuing costs of sequencing (for expendables such as reagents) fell exponentially over most of the time period covered by figure 6.2.[10] When the initial stage of the Human Genome Project was completed in 2001, the authors announcing that feat stated that by 2000 the total costs of sequencing had fallen by a factor of 100 in ten years, with costs falling by a factor of 2 approximately every eighteen months.[11] The cost of gene synthesis is falling at a similar rate. With the caveat that there is only limited data to date, it does appear that the total costs of sequencing and synthesis are falling exponentially (figure 6.4).

The trends of successive shifts to new technologies and increased capability at decreased cost are likely to continue. Using the simple metric of productivity shown in figure 6.2, in the twenty-five years that commercial sequencers have been available, the technology has progressed, from labor intensive gel slab–based instruments to highly automated capillary electrophoresis– based machines to the partially enzymatic pyrosequencing process. Looking ahead to the future of sequencing and synthesis requires first understanding the technology that has produced the progress we have experienced thus far.

A Short History of Sequencing Technologies

The human genome was sequenced primarily using variations of a technique called Sanger sequencing. These methods are based on the fact that DNA molecules of different lengths move at different speeds when exposed to an electric field, a technology called electrophoretic separation.

The Applied Biosystems 377 sequencer, which employed large-format gels to separate DNA fragments of different lengths, was the workhorse instrument of early mass-sequencing efforts. This instrument was labor intensive, as the gels had to be carefully prepared and loaded by hand, a process not amenable to automation.

A significant productivity enhancement came in the form of the Applied Biosystems 3700 and similar machines, in which the gel was replaced by an array of fine glass capillaries. For a variety of reasons, including both basic physics and ease of automation, capillary-based sequencers offered a significant improvement in throughput per instrument. Preparation and sample loading of capillary-based systems can be automated not just for individual instruments but also for arrays of instruments, which led to the construction of sequencing farms at numerous locations throughout the United States, Japan, and Europe. Each sequencing farm produced massive amounts of data.

In these large facilities, laboratory robots perform many of the tasks required for sequencing, and each of these tasks is generally carried out by specialized, dedicated instruments arranged in assembly lines. The arrival of the capillary-based sequencers heralded progressively more-sophisticated technology and ultimately reduced the projected time and cost of the Human Genome Project. Thus fundamental changes in technology reduced human labor requirements, improved throughput per instrument, and through the addition of sophisticated robotics dramatically altered our ability to read genetic information.

A more-recent technological addition is pyrosequencing. Whereas Sanger sequencing relied on enzymatic processing for preparation of samples *before* sequencing, pyrosequencing explicitly relies on enzymatic processing of DNA *during* sequencing. Pyrosequencing has been available in relatively large-scale format sequencers since 1999. A microscale implementation of pyrosequencing was recently used to rapidly sequence whole microbial genomes with an accuracy of greater than 99.99 percent, an important benchmark for the technology before its application to human genomes, starting with Project Jim.[12]

The increased throughput possible with new generations of DNA sequencers is helping to maintain the trend of exponential decreases in cost. There is a constant stream of papers and press releases announcing new

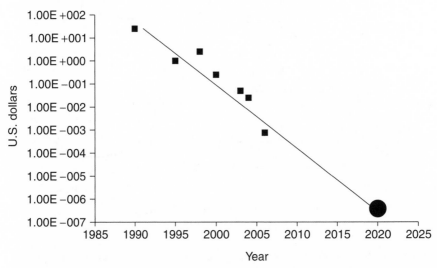

Figure 6.5 Toward the "thousand-dollar genome," a full-duplex human genome (6 billion bases) sequenced for $1,000.

sequencing technologies with ever-greater capabilities, which perhaps will eventually enable low-cost sequencing of individuals and thereby improve health care. However, *at the present rate of improvement,* the thousand-dollar genome—that is, a human genome sequenced de novo for $1,000— may not be available until after 2020 (figure 6.5). New technologies could, of course, provide low-cost sequencing at an earlier date.

How to Write DNA: Synthesis Chemistry

The ability to specify genetic instructions by writing DNA from scratch provides, in principle, the opportunity to specify the behavior of biological systems. The capability to produce synthetic DNA from simpler molecular components is used industrially for many purposes, including production of recombinant proteins, production of gene libraries for protein-selection processes, and synthesis of large DNA constructs. Novel DNA sequences presently can be assembled in two ways. The first method is to harvest molecules extant in nature and then stitch them together into new se- quences; this is the basis of recombinant DNA techniques used for the last thirty years. The second method is to chemically synthesize DNA, in which the sequence is determined by the order in which bases are added to a string of nucleotides attached to a solid substrate. Instruments used to write DNA have benefited from changes in format and automation similar

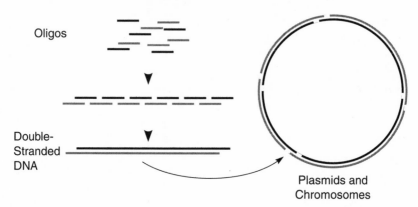

Oligos

Double-
Stranded
DNA

Plasmids and
Chromosomes

Figure 6.6 Assembling oligos into genes.

to those found in sequencing machines, though synthesis chemistry has remained mostly unchanged for three decades.

Unmodified bases from naturally occurring DNA are not yet suitable for
use in existing chemical synthesis methods. Instead, the structure of bases
must be altered by first adding a reactive group in order to facilitate stepwise
addition of each base to a synthetic sequence. The chemical modification is removed during synthesis, leaving synthetic DNA (sDNA) that is biologically
indistinguishable from naturally occurring sequences. Most synthesis instruments are used to directly produce short oligonucleotides, or "oligos," which
are chemically synthesized single-stranded molecules of up to about one hundred base pairs in length. Oligos are then assembled into larger double-
stranded recombinant constructs, such as synthetic genes, and plasmids (see
figure 6.6). Published synthetic constructs now exceed half a million base
pairs in length (see figure 6.1). Commercial DNA-synthesis companies
presently offer delivery of short oligos within a few days, at a cost in the range
of $0.10 to $0.50 per base, and delivery of full genes and longer sequences
within two to four weeks, at a cost in the range of $0.50 to $5.0 per base pair.

Synthesis Technologies

Because the basic chemistry of all DNA-synthesis schemes is virtually identical, performance improvements have come from the parallelization of many
simultaneous synthesis reactions. The first machines built and sold by Applied
Biosystems were capable of synthesizing only four independent sequences at
a time, in relatively large containers (of about one milliliter). The first massively parallel machines were built at the Stanford Genome Research Center
in the mid to late 1990s, in a design later sold through Gene Machines Inc.

These instruments were designed to synthesize oligonucleotides in a 96-well plate format, operating with much smaller reaction volumes. Additional parallelization and miniaturization of plumbing brought the capacity of these instruments to 192 and then 384 independent reaction wells.

Fabricating sDNA on chips exploits a different strategy. Rather than compartmentalizing synthesis reactions in defined containers, an alternative means of synthesizing DNA confines the growing oligos to specific regions on the surface of a silicon chip or glass slide. Synthesis on chips increases speed, reduces reagent usage, and facilitates manipulation of synthesis products. Microfluidic plumbing can be fabricated directly adjacent to the synthesis reaction chambers, thereby reducing human handling and spillage. Some methods used to define the sequence of an oligo at each location are photolithography, local control of acid chemistry via light or heat, and ink-jet printing.

Affymetrix produced the first commercial DNA chips, which were manufactured using photolithography. Each layer (each base in the sequence) requires a specific mask, usually produced with a chrome layer on a quartz substrate; each mask costs approximately $1,000. Thus a production run to fabricate a chip full of oligos fifty bases long could cost as much as $50,000, unless masks can be reused.

Other means may be used to control the local chemistry on the surface of a chip. Rather than use masks with clear and opaque areas to control exposure of the chip to light, NimbleGen makes use of an array of micromirrors that direct light to specific areas on the chip. The micromirror array can be dynamically controlled via a computer, which significantly reduces the up-front production costs. Moreover, it is conceivable that a desktop DNA printer could be designed using this technology, an innovation that would dramatically increase access to synthetic oligos.

Inexpensive ink-jet printers can be used to place bases where desired on the surface of a chip. This technology is primarily commercialized by Agilent. Printing of DNA arrays using pins to spot fluids in precise locations is also possible, though this technique is generally used to print presynthesized oligos, produced either chemically or amplified from an organism via PCR.

Chip-based DNA synthesis has the potential to provide many more oligos per run, at a lower cost, than macroscopic synthesis in multiwell plates. For example, the micromirror array used by NimbleGen contains just over three-quarters of a million mirrors. In principle, each mirror could be used to control the sequence of an oligo. As it becomes possible to manipulate these oligos on the chip, much-longer DNA sequences can be fabricated at lower cost. A first step toward implementing this approach was recently demonstrated by assembling commercially purchased oligos within a microfluidic environment.[13] The researchers demonstrated very low reagent use and a very low error rate. This is clearly just the first implementation of a new and extremely powerful technology, one that will only become more useful as additional sophistication is introduced into the molecular assembly methods.

Biology itself is providing tools to aid in the assembly of large DNA constructs. A recent paper demonstrated the use of a naturally occurring error-correction system, the DNA mismatch-binding protein MutS, to identify mistakes during synthesis.[14] MutS was added to a pool of oligos during assembly. The protein bound to products that contained errors, which were then removed from the pool, thereby enriching the fraction of correct sequences. One round of this procedure produces an error rate to approximately 1 in 4,000 bases (or 1 error roughly every 4 genes). A second round of error correction reduces the error rate to ~1 in 10,000 bases. This rate is so low that a single round of synthesis should be sufficient to produce multigene-length constructs suitable for use in complicated genetic circuits. This technology is already being used commercially and has been extended to the assembly of multiple sequences in the same tube.[15] The inevitable combination of repeated MutS enrichments with microfluidic, on-chip manipulation of DNA-synthesis products will be just the first example of hybrid devices that exploit both emerging biological tools and mature microfabrication technologies.

Anticipating the Future of Synthesis and Sequencing

Accelerated development of sequencing technologies is the focus of large government programs in the United States. And with the explicit goal of bringing the health care benefits of genome sequencing closer to fruition, the X-Prize Foundation in late 2006 launched the Archon X-Prize for Genomics, with $10 million promised to whomever could provide a technology capable of sequencing one hundred people in ten days at no more than $10,000 per genome. By the time this book hits the shelves, that competition may well be over.

Next-generation sequencing instruments employ a great variety of technical strategies, from elaborations of Sanger sequencing to direct identification of bases by their chemical structure. There are so many different strategies, in fact, that exploring the specifics of each is not necessary for thinking about where things are going.

Pushing down the cost of sequencing is the goal of many companies hoping to grab a share of the potentially enormous market for medical genomics. At the time this chapter was written, the race for the thousand-dollar human genome was in full swing. Illumina is presently marketing an instrument capable of providing a billion bases of sequence information per day. Helicos BioSciences promises an even lower-cost instrument that will roll out with a throughput of several hundred million bases per day, expected to be upgradable beyond several billion bases a day as hardware, software, and biochemical processes improve.

If the cost and productivity trends described above continue for another decade, a single person at the lab bench could soon have the capability to sequence or synthesize all the DNA describing all the people on the planet many times over in an *eight-hour day,* even given profligate human reproduction. Alternatively, one person could sequence his or her own DNA within seconds.

Despite the fantastic nature of these numbers, there is no physical reason why sequencing an individual human genome should take longer than a few minutes. Sequencing a billion bases in a thousand seconds would require querying each base for only a microsecond, which is well within the measurement capability of many physical systems. Inexpensive disk drives, for example, already read the state of magnetic domains at upward of a billion times a second. Although storage media is an example of a mature technology, it is also an indication of the sort of interaction that will be possible with biological systems. Indeed, it seems unwise to assume arbitrary limits on potential applications of our newly developing ability to manipulate and probe matter at the scale of individual molecules. Every week there are exciting new examples of imaging and manipulation of molecules or small objects such as carbon nanotubes, each innovation pushing back previously imagined limits. Hybrid techniques utilizing physical measurement of the activity of individual enzymes may provide extremely rapid sequencing.[16] Yet at some point, despite ever-increasing speed, sequencing capabilities will likely reach a plateau in *utility*—how fast is fast enough? This raises the question of how much longer the effort put into developing rapid sequencing technology will be a wise investment.

The greater challenge is sensitivity—biology comes in units of single cells, which is the level we must work at to reprogram biological systems and deal with many diseases. Cancer is one such disease. Generally it is not a whole organ or tissue that becomes cancerous, but rather one cell that breaks loose from its developmental pathway—due to chance mutations, changes in the environment, or infection—and runs amok. Similarly, many infections essentially begin with attacks by individual pathogens on individual cells, even if many simultaneous such events may be necessary to produce full-blown symptoms. The scale of these events is of obvious interest to scientists and clinicians concerned with isolating and understanding novel pathogens, both natural and artificial. Yet no technology presently available commercially can sequence the genome of a single cell without amplification steps that introduce significant errors (though many academic labs and companies promise the ability to do so soon). Most current technologies, particularly those applied to determining interactions between proteins, require a large number of cells and thus produce data that is an average of the states of those cells. Investigating the metabolic or proteomic state of cells (without the use of genetic modification) is similarly often predominantly limited to large samples.

Regardless of the direction of technological development, the synthesis and sequencing capabilities available to an individual in the next decade will be impressive, greatly facilitating the task of manipulating biological systems. The cost of each instrument should generally decrease, following the trend of similar commodities, suggesting that the infrastructure of biological technology will be highly distributed. One indication of this trend is that the parts for a typical DNA synthesizer—mostly plumbing and off-the-shelf electronics—can now be purchased for less than \$5,000. The assembly effort and monetary sum are similar to that expended by many car and computer hobbyists.[17]

Despite existing infrastructure that provides for downloading sequences directly into a synthesizer, possession of a DNA synthesizer does not a new organism make. Current chemical synthesis produces only short runs of DNA. Ingenuity and care are required to assemble full-length genes, although the techniques are already described in the scientific literature. There is significant economic motivation to make such assembly routine, and multiple companies have been founded to sort out the relevant manufacturing details and to take advantage of the growing demand for long synthetic-DNA sequences.

The Geographical Distribution of Technologies

At present, it is more economical for many users to purchase DNA-synthesis services on an outsourced basis than to operate synthesizers inhouse. That is, most people already get their DNA by mail order. Gene-length oligos are already a commodity product, as indicated by the global distribution of commercial DNA-synthesis foundries (figure 6.7). However, mail order of longer organism-length sequences is, thus far, rare. Commercialized methods to assemble the many millions of contiguous bases required to specify bacteria have been labor intensive and therefore expensive. Yet this barrier is certain to fall as automation and on-chip synthesis become more capable or as additional biological tools are included in synthesis. Synthesis companies expect prices to continue to fall rapidly for the foreseeable future.[18]

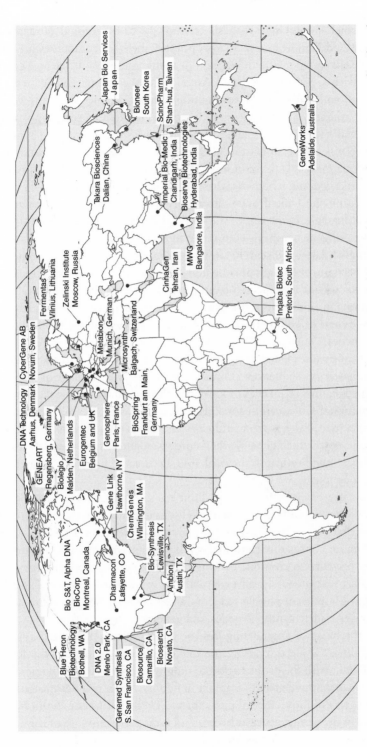

Figure 6.7 Commercial gene-synthesis providers, circa November 2005. Companies that provide gene-synthesis services are distributed globally.

Whether employing synthesis technology of traditional (and publicly available) design or a new and proprietary method, companies are advertising delivery of synthetic genes in a many countries. In 2005 I collaborated with Gerald Epstein and Anne Yu at the Center for Strategic and International Studies (CSIS) to produce a map of commercial gene foundries, which I initially published on the web.[19] A version of the map was later published by *Wired* magazine with some additional data and reporting; figure 6.7 is the version from *Genome Synthesis and Design Futures: Implications for the U.S. Economy.*[20] A more-recent tally of firms providing gene- and genome-length–synthesis services places the number of companies within the United States at twenty-five, with an equivalent number spread throughout the rest of the world.[21] A few comments about the map: the lack of gene-synthesis companies in South America is probably due to the English language search; similarly, companies in Asia and some areas of Eastern Europe might also be underrepresented. While an increased use of oligos would be expected from the proliferation of synthetic-DNA applications in both diagnostic assays and basic science, I remain surprised at the extent of commercial supply around the world for full-length synthetic genes.

The companies listed in figure 6.7 provide synthetic genes via mail; the list is based on sequences submitted over the web, not all of which are screened against sequences of known pathogens and biological toxins. The International Consortium for Polynucleotide Synthesis (ICPS), composed of concerned citizens from industry and academia, recently suggested a framework in which gene-synthesis orders might be monitored to improve security.[22] I will return to the probable success and longevity of such plans in later chapters.

Even if great care is taken to limit the commercial synthesis of DNA from pathogens or toxins, it is unlikely the chemical tricks and instrumentation that companies develop in the course of building their businesses will remain confined within their walls. I believe there will be strong economic pressure to continue commercialization of high-productivity oligosynthesis instruments, and eventually to market instruments capable of accepting inputs of simple reagents and electronic specifications and returning complete synthetic genes and genomes.

At the heart of engineering lies the ability to tinker and to rapidly try possibilities, while discarding the unworkable and pursuing the feasible. Coincident with the proliferation of rapidly improving technologies underlying the economic growth implicit in figure 6.7 is a proliferation of skills. The growth of international participation in the iGEM competition, as explored in the next chapter, is by itself an impressive measure of the appetite for biological technologies around the world.

CHAPTER SEVEN

The International Genetically Engineered Machines Competition

Yᴏᴜ ᴍɪɢʜᴛ ɴᴏᴛ ᴛʜɪɴᴋ that making bacteria smell like bananas could change the world. But over the summer of 2006, a team of five MIT undergraduates, with no prior laboratory experience to speak of, built a genetic circuit that will change the way people think about engineering biological systems.

The bacterium *E. coli* lives in the intestines of mammals and constitutes an important component of the digestive system. Naturally occurring *E. coli*—no surprise—smell like feces. The odor is due to indole, a compound produced by the bacteria and used in intercellular communication and biofilm formation.[1] The five undergraduates, who were participating in the International Genetically Engineered Machines (iGEM) competition after finishing just their first or second years at MIT, embarked on reprogramming *E. coli* odor as a summer project. Several of the students had yet to take an introductory course in genetics.[2] The progress the team achieved over just a few months surprised even their faculty advisors.

As I will describe in more detail over the course of this chapter, the success of the "Eau d'E. coli" project is in no small part due to the application of standard engineering principles to biology. Like modern airplane and computer design, the resulting circuit was assembled not with the the knowledge of many molecular details but with abstraction of those details through the use of parts with predefined functions providing for substantial simplification in the process of design and construction. Only when necessary did the team delve into the details, such as when they developed new parts according to the standards set for iGEM. The synthetic circuit that resulted from the Eau d'E. coli project contained a remarkable twenty-four components.

It is perfectly reasonable to question how substantial any iGEM project is as a step toward true biological engineering. The important point is that Eau d'E. Coli is *a step* forward, though just one of many to come. Furthermore, it as a particularly clear example of engineering methodology applied to biological circuits. Nonetheless, a great deal more progress will need to be made before synthetic biology has a substantial technological or economic impact.

The iGEM was first organized in 2004, and alumni already number well into the thousands and are spread widely across the globe. Each participant has a chance to see the fruits of applying to biological systems an engineering methodology based on composable parts. Each participant has a chance to play with DNA and to build a living system that has never before existed. Most projects fail, and many are overambitious, but every year a good time—and, more important, much education—is had by all. The best way to understand the remarkable progress represented by iGEM is to simply start at the beginning.

THE inspiration for iGEM reaches back to 1978, to a circuit design course taught by Lynn Conway at MIT. That course spread to more than one hundred other universities in the United States within two years. In combination with a textbook by Lynn Conway and Carver Mead, the course rapidly brought very large systems integration (VLSI) circuit design to thousands of students across the country and soon across the world.[3]

The proliferation of VLSI was primarily due to the decoupling of design from fabrication. Prior to 1978, the *electronic* layout of a circuit was dependent on the processes and instrumentation used to produce the *physical* circuit, which meant that a circuit designer had to use design tools specific to each manufacturer. The genius of the Mead-Conway approach was to construct design rules for circuits that could be easily learned and used by engineers and that could also be easily implemented by manufacturers regardless of their particular instrumentation and recipes. There was initial resistance within the industry to this change, with manufacturers insisting that effective design depended on intimate knowledge of silicon processing.[4]

Mead and Conway simply ignored such objections. Conway taught the class, and students enthusiastically adopted the methodology. Students often had working versions of their chips within weeks of submitting their designs.

Fundamentally, the successful implementation of VLSI relies on an "abstraction hierarchy" in which decisions at different levels of design and fabrication become independent. The particular hierarchy utilized by Mead and Conway can be broken down this way:

1. Chip materials and processing techniques are standardized at the level of *atoms and chemistry*.
2. Those materials and processing techniques enable the fabrication of *parts* with defined characteristics, for example, resistors, capacitors, and transistors.
3. Those parts can be used in the construction of simple *devices,* in particular, the logic and memory elements.
4. Those devices form the building blocks of *systems,* such as shift registers, relays, and other functional elements that serve as the building blocks for integrated circuits. Simple systems are to this day packaged as independent integrated circuits for use in building prototypes and in producing everything from appliance-control circuitry to video cards.
5. The final abstraction employed by Mead and Conway was to formulate their circuit design rules to be independent of the actual size of the final object. In 1980 the state-of-the-art technology for building integrated circuits produced wires and other features that were about three micrometers wide. They chose to construct a design methodology that could scale the size of the circuit with a gradual reduction in line width, which is today about forty nanometers, a reduction by a factor of almost 100 over twenty-five years.

Smaller line widths allow for more parts in a given area on a chip, and the resulting increase in density of transistors is the technological underpinning of all the computer power that surrounds us today. The complexity of today's circuitry is possible in large part due to the VLSI design methodology, which through its integration of all the layers of abstraction described above enables standardized parts, devices, and systems to work as expected regardless of the particular manufacturing process. Computer-aided design packages can be used to lay out circuits without ever thinking about the particular atoms or chemistry that will be employed in fabrication. This is precisely the same kind of abstraction, introduced in Chapter 2, that allows cars to be built from parts with defined functions regardless of who fabricated the parts and what materials were used.

In summary, then, Mead and Conway's decoupling of design from fabrication relied on an abstraction hierarchy that let engineers focus on whatever level of complexity they were interested in, without needing to worry about the details at other levels of complexity. Before VLSI, the requirement to consider the specifics of fabrication made circuit design extremely cumbersome and complex. VLSI refocused designers from thinking about atoms and process chemistry to thinking about transistors and circuits, thereby completely transforming the industry.

Within two years, chips designed using the VLSI rules appeared on the market, beginning in 1980 with the Motorola 68000. The architecture of that chip is still in use, and the progress toward a design-to-build methodology (see Chapter 5) that culminated in the contributions of Mead and Conway has been refined further still. Today, producing a new chip remains a matter of sitting down with a computer, designing circuitry in software, sending off the design to a fabrication plant that could be anywhere in the world, and waiting for the final product to show up via the mail. Those chips now contain of many billions of components. This style of engineering is the foundation for hundreds of billions of dollars in annual revenues worldwide, from products ranging from the hardware that supports the Internet to cell phones to computers that are ever more powerful and ubiquitous.

Bringing Complexity Engineering to Biology

The cumulative effect of standardization, decoupling, and abstraction has been to take the "electrical" out of electrical engineering. Tom Knight, a senior scientist at MIT's Computer Science and Artificial Intelligence Laboratory and an early participant in designing hardware and software for ARPANet, sees his field now as "complexity engineering": "In most respects the association of electrical engineering with electromagnetism is now almost incidental. We have become complexity engineers, rather than experts in [electricity and magnetism]. Few other disciplines design, build, and debug systems as complex as a modern computer system, either from the perspective of hardware, with billions of components, or with software, with millions of lines of code."[5]

The utility of applying methods developed for use in computer system design to biological system design was apparent to Knight early on. The likelihood and potential consequences of applying complexity engineering to biological systems *successfully* is another matter, of course. Nonetheless, Knight is forging ahead, trying to make as much progress as he can. In the early 1990s, he began teaching himself molecular biology and genetics, with the aim of learning to reprogram organisms to build new objects at the molecular scale. Based on his experience designing and building computing systems from the ground up, the role of complexity engineers in creating a new approach to biological engineering was, to Knight, crystal clear: "We have an opportunity to use our complexity and information management tools to . . . modularize, abstract, and understand biological systems. In the same way that we simplified and abstracted components from physics to allow us to build billion component processors, we can and will modularize, abstract, and understand biological components with the explicit goal of constructing artificial biochemical and biological systems. I believe no other discipline can effectively do this."[6] Another way to describe this capability is "forward engineering" or "forward design," the use of models of parts with understood behaviors as the basis for constructing more-complex objects. Consequently, Knight began his project with (1) the development of a set of biological parts with defined functions and (2) a method to easily assemble those parts into devices. In 2001 Knight was joined at MIT by Drew Endy (see Chapter 4) and Randy Rettberg, an engineer and a former executive at Sun Microsystems and Apple, in founding the Synthetic Biology Group at MIT.

In 2003 the group initiated an independent activities period (IAP) course in synthetic biology at MIT, explicitly modeled on Lynn Conway's efforts of a quarter century earlier. The commercial fabrication market for synthetic genes was already emerging at that time (see Chapter 6 and figure 6.7). Knight and his colleagues had to begin assembling the rest of the engineering tools that would enable building biological circuits in ways analogous to those used by Conway's students. In 2003 neither the set of tools nor the set of parts was sufficiently complete to build more than rudimentary genetic circuits. In fact, most groups were able to accomplish only construction of models and initial assembly of their circuits. The entire project was at such an early stage that "during the 2003 course students helped to invent and implement ideas about 'standard biological parts.'"[7]

Those students were sufficiently satisfied with the experience that the instructors repeated the course in 2004.[8] The IAP courses served as an initial demonstration of design and assembly methods, and the students contributed a small number of parts to the Registry of Standard Biological Parts (introduced in Chapter 5). Some of the parts used most frequently in subsequent years were initially designed and characterized by the IAP students.[9] In late 2003, the Synthetic Biology Group was funded by the National Science Foundation, to expand the course to five schools. In the summer of 2004, iGEM was launched with the participation of Boston University, Caltech, MIT, Princeton University, and the University of Texas at Austin. (The *i* in iGem initially stood for "intercollegiate," and the first year all the participants were graduate students.)[10]

Before we proceed to a description of the accomplishments of the participants, it is worth revisiting the abstraction hierarchy that enables building
anything with interchangeable biological parts. While many different choices
could be made about how to define parts, in what follows I hew to the definitions promulgated by the BioBricks Foundation and used by iGEM:

1. The *atoms and chemistry* constituting BioBrick parts and their assembly methods are presently constrained by how preexisting organisms
   encode information and construct molecular machines.
2. DNA serves as the material with which *parts* are fabricated. Func-
   tionally, a part could be green fluorescent protein (GFP), yellow fluorescent protein (YFP), or a DNA-binding protein that serves as a
   repressor, as described in the examples of circuits in Chapter 4.
3. When combined with an inducer signal, such as heat or IPTG (the
   chemical used in the switch and repressilator discussed in Chapter 4),
   parts can be used to build a *device*, such as an inducible promoter that
   allows control of gene transcription from outside the cell.
4. As demonstrated by the genetic switch and the repressilator described
   in Chapter 4, combining *devices* in different ways allows construction
   of *systems* that constitute genetic circuits.
5. At present the closest analogy to the scale-free design rules adopted
   by Mead and Conway may be the requirement that DNA delivered by
   fabrication houses be the sequence that is ordered. There are not yet
   very many ways of producing synthetic genes (see Chapter 6), but as
   long as a particular fabrication house can deliver a given sequence,
   then the details of synthesis are irrelevant.[11]

So, after this optimistic and enthusiastic outline of the methodology,
what did the students come up with? An interesting part of the history here
is that records of the first few years are already spotty. As with the rest
of biology, much of the detailed information about iGEM presently resides solely in the memories of the participants. Not all of the early teams
contributed materials that described experimental results, a requirement
later instituted to help develop the memory of the community. Some early
projects resulted in publications or were written up as part of university
course requirements for participating students, but uniformly accessible
records are not available for years before 2006. In addition, the early years
of iGEM were often plagued by a desire for synthetic genes that was not
commensurate with the industry's actual capabilities; many teams did not
receive the genes they ordered in time to complete their projects for the
competition.

In the sections that follow, I delve into detail that may be beyond the interest of all readers. The last section of this chapter returns to a more-general discussion of iGEM in the context of historical technology development. For the sake of brevity, I describe below only selected projects among those that achieved substantial, and documented, results. My intent is to convey a general sense of what undergraduates and high school students have, and have not, accomplished for summer projects and thereby provide a context for grasping how quickly the field is changing. By the time iGEM 2007 concluded, many of projects had already achieved a level of complexity that put the details of construction and operation well beyond the scope of this book.

iGEM 2004

First Bacterial Photograph

The original goal of the team from UT Austin was to engineer a biofilm of bacteria that could detect edges in projected images. The resulting pattern in the bacteria would consist of lines showing transitions from light to dark in the original image. This capability could, for example, be used to pattern the synthesis or deposition of materials simply using light, thereby achieving one of Tom Knight's original goals in applying complexity engineering to control the biological production of objects.

The team demonstrated a system that allows for the control of gene expression in *E. coli* by using light.[12] The circuit turns off the production of a black pigment when illuminated, and a mask can be used to determine the specific pattern of light projected onto the bacterial film. The demonstration was by itself such a sufficiently novel and interesting result that it was published in *Nature* in 2005.[13]

E. coli do not naturally possess photoreceptors, so the first order of business for the UT Austin team was to create that capability. After trying different approaches, the team was fortunate to find that an entire light-controlled expression device had been constructed earlier in the year. The first graduate student in the lab of Christopher Voigt at the University of California, San Francisco, Anselm Levskaya, built the device for his first project and later donated it to the UT Austin team.

Neither Levskaya nor Voigt had heard of iGEM when they decided to pursue the project.[14] Levskaya used as components naturally occurring genes found in organisms all over the planet and stitched them together using standard tools from molecular biology. The end result was a new phototransducer protein, stitched together from a cyanobacterial light sensor and a salt sensor from *E. coli* that modulates gene expression in response to red light.[15]

It is interesting to note here that the specific physical details of how the phototransducer actually converts light into a molecular signal are still rather unclear. That is, the story remains largely unwritten about what happens *between* the time when a photon is absorbed by the receptor and the time when the transcription factor is modified. While the original cyanobacterial photoreceptor is understood, and the original *E. coli* salt sensor is understood, there is no device physics description of the new part, a chimera that exists only through human effort in the laboratory. Moreover, rather than being the product of forward design based on quantitative models of the constituent components, the new phototransducer was produced by first manufacturing many different versions and then selecting the one that worked the best.[16] Nonetheless, the input and output of the part have been well characterized experimentally. The existence of this detailed data, rather than an atomic-level, mechanistic understanding of the protein, constitutes a sufficient description of the new part to enable its use in rational design of devices and systems.

RNA Caffeine Sensor

The team from Caltech departed from the abstraction hierarchy in an interesting way. Rather than relying on proteins as functional elements, the Caltech team used RNA.

Recall from the discussion of the Central Dogma of molecular biology (Chapter 4) that nucleic acids were for many years thought to have no function beyond information storage and transfer. As with many of the stories told about biology, this one had to be amended as experiments demonstrated that RNA could, in fact, perform functions similar to those of many proteins. In particular, just like proteins, some RNA sequences can fold into complex structures and bind small molecules, in this case, caffeine. The Caltech team combined this and other RNA technologies into a set of molecules that could distinguish between three levels of caffeine in coffee.[17]

iGEM 2005

The year 2005 brought the first international participants to iGEM, with teams from Toronto, Canada; Cambridge, the United Kingdom; and Zurich,

Switzerland. Total participation consisted of one hundred students making up thirteen teams from four countries.[18]

The year 2005 also brought the realization that the students' appetites for designing new parts, devices, and systems outstripped the existing technical ability to construct all those things. A news piece in *Nature* noted that "if none of the designs succeeded completely, that was more because of the limitations of the nascent science of synthetic biology than any lack of enthusiasm, creativity or hard work."[19]

The main stumbling block in 2005 remained assembly of the parts into functioning circuits, in large part because synthesis companies at that time were still struggling to successfully provide gene-length sequences. Beyond fabrication difficulties lay more-complex issues of design and simulation. The team presentations still consisted largely of complex, colorful circuit diagrams and simulations but little data, demonstrating once again that synthetic biology was very young in comparison to the VLSI infrastructure the iGEM organizers were trying to recapitulate.[20] The teams eventually contributed hundreds of new parts to the Registry of Standard Biological Parts, though many have not been fully tested and characterized. Drew Endy noted at the time, "We don't know how to engineer biological systems. You can't teach something you don't know how to do, so the students are helping us to figure it out."[21]

iGEM 2006

In 2006 the competition expanded to include four hundred students from thirty-eight schools in fourteen countries.[22] This year also brought the idea of a competition to the fore, with prizewinners in various categories. The grand prize-winning team hailed from the University of Ljubljana in Slovenia. The first runner-up team was from Imperial College in London, and the second runner-up was from Princeton University. Recognized categories included best part, best device, best system, best presentation, and, explicitly acknowledging the continuing difficulty of getting anything at all to work, best conquest of adversity.[23] Additional prizes were occasionally created spontaneously by the judges. While many teams were able to produce and characterize new parts and devices, very few completed the construction and testing of new systems. Communication of results improved dramatically, however, as almost all the teams submitted their results in the form of presentations, now archived on the iGEM website.[24] Papers describing many of the teams' results were published in an online journal.[25]

Engineered Immune Response

The team from Ljubljana decided not to work directly on bacteria but to build a system in human cells that could aid in understanding what happens during infection.[26] The target of the project was the response of the human immune system to pathogens, a response that in some cases can get out of control and become exaggerated. This, in turn, can lead to over-inflammation and fever—symptoms of sepsis, which kills at least thirty-four thousand people annually in the United States and afflicts as many as 18 million people per year worldwide, with an overall fatality rate of about 30 percent.[27]

Programmed Differentiation of
Embryonic Stem Cells

The Princeton team had the modest goal of developing "reliable techniques for programmed tissue generation in mammalian systems."[28] Differentiation of stem cells into functional tissues is a poorly understood process; just as in building airplanes, one way around trying to understand the complexity of the naturally occurring control system is to substitute an engineered system for the one that we have inherited.

As part of the effort, the team began the construction of an all-mammalian stem cell BioBrick registry containing parts suitable for use with a viral delivery system. At the time of the presentation, the team had produced seventy functioning mammalian BioBrick parts. Like the 2005 bacterial photography system of UT Austin, this project was sufficiently complex to require several years of effort to meet the original design goals. Publications describing the work are presently in preparation.[29]

An Arsenic Bacteriosensor for Drinking Water

Students from the University of Edinburgh took home the inaugural prize for best real world application for their attempt to construct an arsenic detector for drinking water. Arsenic poisoning is thought to affect up to 100 million people worldwide. An estimated 20 million people in Bangladesh alone drink unsafe water, from wells inadvertently drilled through an arsenic-rich band of sediment.[30] Existing assays for arsenic in drinking water require sending samples to laboratories for expensive testing, which return a high number of false negative results. Moreover, existing assays have a sensitivity limit of 50 parts per billion (ppb), well above the World Health Organization's recommendation of no more than 10 ppb.

In an effort to provide inexpensive field-testing for arsenic, the students from Edinburgh built a synthetic detector system in *E. coli*. While arsenic itself is hard to directly detect chemically, engineering a genetic switch triggered by arsenic could enable a bacterium to generate a signal that is easily measurable using simple pH-sensitive dyes.

The students demonstrated that the sensor could detect arsenic at concentrations as low as 5 ppb. Following on the work of the iGEM team, a graduate student in the lab of team advisor Chris French has now produced a strain of bacteria carrying the arsenic detector, which can be freeze-dried for easy shipping. Publications describing the work are in preparation, and several companies have expressed initial interest in commercializing the test.[31]

Not-So-Smelly Bacteria

The Eau d'E. Coli team designed two systems that could process endogenous metabolites into the compounds responsible for wintergreen (methyl salicylate) or banana (isoamyl acetate) smells. In order to simplify the design, enable troubleshooting, and provide a means of control in the event that various components failed to work, the team hewed closely to the Drew Endy's abstraction hierarchy.

Both systems were composed of two devices each. The complexity of the design was daunting. For example, the Wintergreen Odor Biosynthetic System was composed of the Salicylic Acid Generating Device (SAGD), which converted the cellular metabolite into salicylic acid, and a Wintergreen Generating Device (WGD), which converted salicylic acid into methyl salicylate. The WGD could be tested by adding salicylic acid to the medium in which the bacteria were growing, which served as a backup in case construction of the SAGD failed. Similarly, the Banana Odor Biosynthetic System was composed of the Isoamyl Alcohol Generating Device (IAGD) and a Banana Generating Device (BGD). The system was designed so that either isoamyl alcohol or 3-methylbutanal (which is naturally metabolized by *E. coli* into isoamyl alcohol) could be added to the growth medium for troubleshooting the BGD.

Both the Banana Generating and Wintergreen Generator Devices worked as designed as soon as they were assembled.

The team managed to successfully build and test the entire wintergreen synthetic system. However, while they also managed to construct the entire banana synthetic system, the Isoamyl Alcohol Generating Device apparently failed in this context; only when the growth medium was supplemented with isoamyl alcohol did the system produce a banana odor. Thus the final design did not quite function as desired, in part because, consistent with the experience of many others, the team ran out of time.

Yet in just four months, a team of MIT undergraduates managed to make substantial progress on an engineering task that just a few years earlier would have posed a serious challenge to a team of professional molecular biologists. Their success was due in large part to the existence of the Registry of Standard Biological Parts and the emphasis of iGEM organizers on working within an abstraction hierarchy. The description of this work as building wintergreen- and banana-odor generators may serve to trivialize the accomplishment. By no means should the magnitude of the team's achievement be underestimated. The students rapidly prototyped a complex, synthetic metabolic pathway composed of twenty-four parts drawn from four organisms. For all its imperfections, the Eau d'E. coli project soundly demonstrates that forward engineering of biological systems is both feasible and practical.

iGEM 2007

The year 2007 saw 750 participants hail from nineteen countries, making a total of fifty-four teams.[32] I was fortunate to serve as one of about twenty judges for iGEM 2007. The rules for participation and scoring in 2007 included the requirements to submit new parts and devices to the registry by the time of the iGEM Jamboree, document (though not fully characterize) those parts, and maintain a website describing models, designs, methods, and experimental data.

The finalists included two teams from the United States, two from Europe, and two from China. The grand-prize winner, and new custodian of the BioBrick trophy, was the team from Peking University, in Beijing, China. The other finalists included teams from the University of California, Berkeley; the University of California, San Francisco; a coalition of institutions in Paris, France; the University of Ljubljana, Slovenia; and the University of Science and Technology of China Beijing.

Control Circuits for Bacterial Differentiation

The grand prize–winning project from Peking University demonstrated new devices to control the spatial and temporal activity of bacterial cultures. The overarching goal was to create the capability to control the behavior of a large number of cells by giving them the ability to "differentiate out of homogeneous conditions into populations with the division of labor."[33] In other words, the Peking team wanted to lay the groundwork for programming complex behaviors into groups of genetically identical single-celled bacteria.

Synthetic Multicellular Bacterium

A multi-institution team from Paris set out to build a set of components in *E. coli* that would turn the single-celled bacterium into a multicellular organism.[34] As in the case of the project from the Peking team, the objective was to enable the programming of complex production systems. However, the Paris team choose to pursue the engineering of complexity not through building increasingly complex circuits within a single cell—which might result in unanticipated, and undesirable, interactions as complexity increased—but rather through building modular complexity with simpler circuits in different cell lines. Separated physically by cell membranes, the specialized cell lines were intended to interact in well-defined ways to produce complex phenomena.

Extensible Logic Circuits in Bacteria

The team from the University of Science and Technology of China (USTC) sought to develop new, independent molecular signals for use in genetic circuits. Recall that the genetic switch and repressilator (Chapter 4) used naturally occurring repressor proteins as elements to regulate transcription of DNA into RNA in an artificial circuit. The number of preexisting, independent bacterial repressors easily co-opted for use as parts is small, which limits the number of different control algorithms in which they can be included.

The USTC team sought to remedy this situation. Through both theoretical design and experimental searches, using mutation and evolution, the team produced a large number of synthetic repressor proteins and other components.[35] (During its oral presentation, the team noted somewhat dryly, "The experimental approach is really time consuming." Indeed.)

Location, Location, Location

The team from University of California, San Francisco, built two systems to spatially control the flow of information and materials within cells, a project they called "Location, Location, Location."[36] This team deserves special attention because, as UCSF has no undergraduate program, the team consisted high school students from San Francisco and Palo Alto.

The team built two systems. The first was related to a project in the laboratory of Wendell Lim, who studies how signal transduction pathways are constructed using protein complexes. A common feature of such pathways is a large "scaffold protein," a backbone with binding sites for other members of the pathway. Collecting pathway members together on a scaffold is one way to organize both the order and the rate of reactions by controlling the physical proximity of components. The team was able to add a module to an existing pathway in yeast that provided a new "knob" to control the output of the pathway.

The second system consisted of the machinery necessary to build a "synthesome," an artificial organelle. This is an isolated compartment that would, in effect, be a microenvironment that might be used as a bioreactor to produce drugs or fuels. Moreover, the team conceived of a way to label the synthesome with a particular molecular tag that could be used as a unique identifier within the cell. This tag could then enable the specific targeting of metabolites or protein complexes to the synthesome. The team demonstrated portions of the machinery necessary to produce the organelle but did not have sufficient time to demonstrate complete assembly or to show specific targeting of metabolites or protein complexes.

Bactoblood

The team from UC Berkeley designed and built Bactoblood, a system that produces hemoglobin in *E. coli* and that might someday be used as a replacement for red blood cells in emergency transfusions.[37] The design objective was to produce a blood substitute that not only could be easily stored after being freeze-dried but also could be easily reconstituted and grown in large volumes.

First, the team deleted several genes that code for proteins which cause toxic effects in humans, thereby protecting humans from the bacterium. In an attempt to eliminate the ability of the bacterium to grow at all within a human body, they also deleted the genes that enable *E. coli* to take up and metabolize iron. Second, because the human immune system is extremely good at identifying and destroying foreign matter, the team added a gene that enables *E. coli* to build a protective capsule around itself, thereby protecting the bacterium from its host.

Then the team inserted genes coding for the machinery to produce hemoglobin and to help the bacteria recover from being freeze-dried. Additional circuitry included a plasmid coding for proteins that destroyed the bacteria's genome but left the cell physically intact. The self-destruct circuit and the hemoglobin-production machinery are both controlled by switches.

Thus the bacteria could be reconstituted and grown to large volumes, and external signals would simultaneously start hemoglobin production and begin the process of destroying the genome. This combination was supposed to render the bacterium an inert bag of hemoglobin.

To say that this project is a long way from clinical trials is, of course, understating the amount of work necessary before anything like Bactoblood might be used in humans. Nonetheless, the team built and demonstrated all these components, showing freeze-dried samples of the system during the Jamboree.

Virotrap

The team from Ljubljana designed two pathways to prevent infection of human cells by HIV, the virus that causes AIDS.[38] Because HIV mutates often, the Slovenian team wanted to create protective mechanisms that were not specific to a particular viral sequence but that interfered more generally with viral function. The defense mechanisms, called Virotrap, would be activated only when HIV bound to target cells, after the mechanism detected either (1) the binding of HIV to the extracellular portion of proteins embedded in the cell membrane, a prerequisite to infection of a host cell by the virus or (2) the activation of specific viral proteins present in the cell after infection. Either signal could lead to transcription of genes that cause apoptosis, or cell suicide, a common natural response to infection, or could lead to transcription of genes involved in the cell's built-in antiviral immune response.

The team successfully built and demonstrated devices that detect HIV binding and that detect the presence of the HIV protease protein. The team also demonstrated devices that, when triggered, lead to cell death or to the initiation of an immune response. The mechanisms by which these devices accomplished their designed functions are extremely clever, and worth understanding, but are also sufficiently complex to be beyond the scope of this book. Similarly, the practical engineering challenge of building and testing the constituent parts was substantial and is beyond the present discussion.

I would be remiss not to observe the potential for integrating Ljubljana's systems with those demonstrated by the Princeton team in 2006. Combining Princeton's control circuitry for programmed differentiation of human stem cells with Virotrap could be a potent tool to use against disease. This approach is obviously not without its risks; genetically modified stem cells were the cause of the X-SCID leukemias discussed in Chapter 2. Nonetheless, while the laboratory demonstrations from Princeton and Ljubljana are a very long way from testing in humans, the progress made by undergraduates in not just *imagining* new uses for stem cells and defense pathways but also in *building* working versions of those systems should give the reader both pause and cause for excitement.

Microbial Biofuels Production

Notable in 2007 was the first appearance of projects aimed at building systems to produce biofuels in microbes. None of these projects proceeded far enough to actually produce fuel, but they did provide to the Registry of Standard Biological Parts many parts that might be employed in future projects. I will return to the promise of this technology in Chapter 11.

Summary: On the Utility and Inevitability of Bringing Complexity Engineering to Synthetic Biology

The lesson to take away from the efforts of iGEM participants is that complexity engineering in the service of synthetic biology, at least as defined by the group at MIT, is in its infancy. Most designs don't work, either because most BioBrick parts are not fully composable or because they are poorly characterized.

Given that iGEM draws inspiration from the VLSI course taught by Lynn Conway, some context from the history of integrated circuits is useful to gauge the progress of students in building synthetic biological systems.[39] Nearly half a century passed between the demonstration of the transistor and the widespread use of that invention in integrated circuits that power the Internet. The iGEM competition and the Registry of Standard Biological Parts have been in existence only since 2004. Only a few of the "standard" parts are well characterized, and many team projects are not completed. That said, the utility of the Registry is already becoming apparent through the reuse in new projects of many parts and previously demonstrated devices.

This vision of biological engineering—of BioBrick parts and quantitative design tools as forming the basis of a synthetic biology—is yet another story, an analogy that may be useful in understanding where technology is going. While the discussion above leaned very heavily on that particular vision, I return in to a more-general discussion of biological engineering in the next chapter.

CHAPTER EIGHT

Reprogramming Cells and Building Genomes

A FROTHY TANK OF MICROBES growing in Berkeley, California, may hold the key to producing an inexpensive cure for malaria. Within that tank, genetically modified yeast churn out the immediate chemical precursor to artemisinin, the most effective antimalarial drug in the human armamentarium. The precursor, artemisinic acid, can be easily converted into several different versions of the drug, which for the sake of simplicity I here lump together as "artemisinin." A single seven-day course of the drug cures the disease. In regions where adherence to weeklong-treatment courses is poor, artemisinin in combination with other drugs reduces treatment to three days. Artemisinin combination therapy (frequently referred to as ACT) is also employed as a strategy to slow the inevitable rise of drug-resistant malaria parasites. Unfortunately, artemisinin is presently too expensive for most affected people or their governments to afford. Even when the patient is cured, malaria infections can recur after subsequent bites by carrier mosquitoes, further driving up the cost of annual treatment.

Jay Keasling, a professor of chemical engineering at University of California, Berkeley, and chair of the new Synthetic Biology Department at Lawrence Berkeley National Laboratory, began a project to produce artemisinin in microbes in 2000. Keasling's group has used twelve genes from three organisms to build a new metabolic pathway in yeast, a task significantly more complex than most contemporary genetic modification efforts.

As pursued by Keasling, drug production in microbes is similar in structure and operation to brewing beer and can be ramped up very quickly, producing drugs within hours of start-up. Microbial production of therapeutic compounds should, as a general strategy, dramatically improve access and reduce costs. In December of 2004, the Bill and Melinda Gates Foundation awarded a five-year, $42.6 million grant to a collaboration of nonprofit and commercial partners to bring Keasling's vision to reality. After reaching the cost and yield milestones specified in the original grant, those collaborators have embarked on a partnership with the pharmaceutical company Sanofi-aventis to commercialize the technology. Keasling expects to reach his goals of producing artemisinin in large quantities at low cost by 2010.[1]

The Economic Burden of Malaria, and the Value of a Synthetic Remedy

Cost remains the primary obstacle to widespread use of antimalarial drugs. Many affected families can spend only pennies a day on drugs, and the per capita health budget in malarious countries often amounts to the equivalent of only a few U.S. dollars per year.

The cost burden of the disease on individual families is highly regressive. The *average* cost per household for treating malaria may be in the range of only 3–7 percent of income, but total and indirect costs to *poor* households can amount to one-third of annual income.[2] The disease also disproportionately affects the young. Approximately 90 percent of those who are killed by the parasite are African children under the age of five; according to the World Health Organization (WHO), a child dies from malaria roughly every thirty seconds.[3]

In addition to staggering personal costs, the disease harms whole societies by severely inhibiting economic development. In affected countries, malaria reduces GDP growth by about 1.3 percent per year.[4] These countries, moreover, contain about 40 percent of the world's population. Over the past forty years, the growth penalty has created a difference in GDP that substantially exceeds the billions in annual foreign aid they receive. In 2000 the World Health Organization estimated that eliminating this growth penalty in 1965 would have resulted in "up to $100 billion added to sub-Saharan Africa's [2000] GDP of $300 billion. This extra $100 billion would be, by comparison, nearly five times greater than all development aid provided to Africa [in 1999]."[5]

Because there was no technical means to eliminate the parasite in the middle of the twentieth century, this is clearly a number calculated to impress or shock, but the point is that the growth penalty continues to balloon. As of 2008, the GDPs of countries in sub-Saharan Africa would be approximately 35 percent higher than they are today had malaria been eliminated in 1965. The World Health Organization reckons that malaria-free countries have a per capita GDP on average *three times larger* than malarious countries.[6] The productivity of farmers in malarious countries is cut by as much as 50 percent because of workdays lost to the disease.[7] The impact of producing an effective and inexpensive antimalarial drug would therefore be profound.

Improving access to other technologies, such as bed nets treated with insecticides, would also be of substantial aid in reducing the rate of infection. Yet infected victims will still need access to cures. Prevention might be found in a vaccine, which the Gates Foundation also funds. However, even the most promising malaria vaccine candidates are only partially effective and cost even more than artemisinin. Microbial production of artemisinin would completely change the impact of malaria on billions of people worldwide.

Artemisinin is presently derived from the wormwood tree and has been used as an herbal remedy for at least two thousand years. Its antimalarial activity was first described by Chinese scientists in 1971.[8] The existence of the drug and its physiochemical properties were announced to the world in 1979, although its precise molecular mechanism of action is still not understood. A method for chemical synthesis was published in 1983, but it remains "long, arduous, and economically nonviable."[9]

Because natural artemisinin is an agricultural product, it competes for arable land with food crops, is subject to seasonal variations in supply, and its production costs are in part determined by the costs of fertilizer and fuel. As a result of the work of Keasling and his collaborators, it appears that, within just a few years, biological technology may provide a more-flexible and less-expensive supply of drugs than now exists. Commercial production of artemisinin should commence in 2010, with a continuous annual production sufficient to treat the 500 million malaria cases per year.[10]

Brewing Therapeutic Drugs

Keasling's approach to engineering metabolic pathways is in principle the same as pursued in many other engineering projects: he started by breaking the problem into smaller pieces that could be solved individually. This is not so dissimilar from the approach used to build cars, airplanes, computers, or even the undergraduate projects for iGEM (see Chapter 7).

However, unlike the bottom-up approach followed in iGEM using composable BioBrick parts, Keasling and his collaborators do not yet have access to individual pathway components with defined device physics. Moreover, given the ongoing human cost of malaria, the team cannot wait for those components to be developed. Team members must invent as they go. In order to address a significant human need, they are faced with constructing a new pathway that is technically *and* economically functional, using whatever tools and components they can bring to bear.

In particular, when the project began, gene sequences for artemisinin pathway components in wormwood were available electronically, but a key gene was not available as physical DNA. Therefore, Keasling's team resorted to adopting a version of that gene from another plant, which they identified within a sequence database. The gene was synthesized from scratch and optimized for expression in yeast (see Chapter 6), and that step alone improved the yield of a precursor to artemisinin by a factor of several hundred.[11]

The choice to use yeast as a host for the synthetic pathway required careful reengineering. Eight of the twelve genes in the synthetic pathway were drawn from yeast. Therefore, constructing a new pathway in yeast required balancing the native use of those genes with the synthetic use. At times, when progress was slowed by the complexity of working in yeast, the team performed development and troubleshooting experiments in *E. coli* and then moved the resulting circuit back into yeast.

As this is an engineering project, one might expect that a mathematical model of the system would be useful in optimizing the various components. Unfortunately, as the first half of this book illustrates, building accurate models to aid in engineering the metabolism of even a microbe is still difficult work. Keasling notes that while modeling the flow of material is sometimes useful as a guide to manipulating central metabolism, he has not found it as useful for engineering linear pathways like the one constructed to produce artemisinic acid.[12] But the team obviously did not let the lack of a comprehensive model of all the system components stand in their way. In addition, unlike the participants in iGEM, the team had no access to a registry of standard parts to draw from, no standardized assembly protocol, and no abstraction hierarchy to aid in design, construction, and troubleshooting.

Metabolic engineering could be loosely defined as exploiting all available tools and methods to build organisms that catalyze or synthesize desired compounds. While the resulting system might be less well understood than one resulting from ground-up construction via Tom Knight's complexity engineering, and while the behavior of the individual parts used in construction might be less well defined than that of BioBrick parts, the object of metabolic engineering is to get the job done regardless.

Keasling's description of this work makes clear both the difficulty and the ad hoc nature of the engineering challenge. Tinkering with pathways has many confusing side effects, and considerable hard work is required to sort out a functioning system.

Keasling's team has had to tailor each and every step in the pathway to function within yeast. As a result, it is unlikely that the pathway can be easily ported to a different host. Moreover, many steps forward were serendipitous; at one juncture in his story Keasling usually comments, "At this point, a number of miracles began to simultaneously happen in the lab."[13] Those miracles were certainly welcome, were perhaps necessary to progress, were possibly understood in retrospect, but by definition were completely unplanned. This again serves to emphasize the differences between biology and fields in which predictive design is better established. And yet despite the limited capability to build according to a design, biological technologies are making substantial contributions to the economy, as we shall see in Chapter 11. Those contributions, however, presently come at the cost of large investments in labor.

In a manuscript describing the Eau d'E. coli project, the Massachusetts Institute of Technology iGEM team from 2006 (see Chapter 7) notes that "[while] metabolic engineers have demonstrated successful construction of novel biosynthetic pathways in industrial microorganisms for the purpose of producing commercially useful compounds . . . such engineering feats require huge investments of labor, time, and capital by world-renowned genetic engineers. Such large resource investments are generally only justified when the product is overwhelmingly compelling from a commercial or health perspective."[14] Keasling's group has already invested over ten person-years of effort into the project, with even more contributed by commercial partners. While all that work has resulted in a roughly billionfold increase in the yield of artemisinin compared to the initial demonstration experiments in 2001 (figure 8.1), economically viable production will require another roughly hundredfold increase.[15]

Synthetic biology already offers a lever to engineers interested in using biology to build systems with less-obvious or less-immediate commercial value. Eventually, however, the ability to compose new circuits using all the tools of complexity engineering could substantially improve the ability to build new biological systems. Keasling, too, looks forward to access to novel synthetic control elements and suggests that synthetic biology, through the use of composable parts, will be critical in reducing the effort now required to build synthetic genetic circuits.[16]

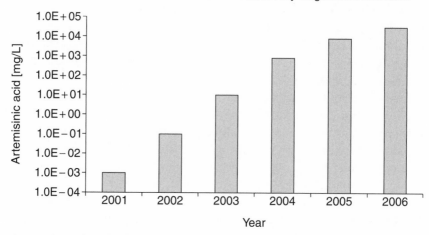

Figure 8.1 The yield of artemisinic acid produced in yeast has increased almost a billionfold in six years. Source: Jay Keasling, UC Berkeley Faculty Forum on the Energy Biosciences Institute, 8 March 2007, http://webcast.berkeley.edu/event _details.php?webcastid=19207&p=1&ipp=1000&category=.

One difficulty inherent in building new circuits on top of those nature has produced is not only that most circuits are not insulated from the metabolism of the host but also that the host itself is usually not well understood as a platform. This has led to a number of efforts to produce organisms with relatively simple genomes that might serve as a programmable host.

The Chassis and Power Supply

If you want to build a computer from scratch, a quick stop at an electronics store will net you a chassis with a power supply. This box has standard fittings for components such as disk drives, a motherboard, or a graphics card, fittings that provide the opportunity to plug in whatever combination of bits and pieces you desire. Chassis come in all shapes and sizes, with power supplies suitable for different mains voltage and frequency standards. Internally, chassis are characterized by different connectors for communication between circuit boards and by the framing and screw holes suited to the physical dimensions of those boards.

E. coli is currently a chassis of choice for many biological programming projects. Unfortunately, even though *E. coli* is a fairly simple organism, it still contains many uncharacterized genes. Of the approximately four thousand genes in the most commonly used laboratory strains of *E. coli*, more than 10 percent still have no functional assignment.[17] One recent effort at reducing this uncertainty compared different strains of *E. coli*—strains that were cultured from different environments and that have different numbers of genes—to identify a smaller common core genome, thereby highlighting "extraneous" genes that do not appear in all strains. The extraneous genes were then deleted from the strain with the shortest genome. This resulted in a new organism, whose genome had been reduced in length by about 15 percent, to just under 4 million bases, and which is now marketed as Clean Genome *E. coli*.[18]

It is not clear what functional changes those deletions effected, in part because many of the protein interactions in *E. coli*, "remain obscure."[19] A major goal of the project was to eliminate genes that lead to mutations and genome rearrangements in response to stress. In addition to the expected useful characteristic of increased genome stability, the strain (ironically) also unexpectedly displayed several other "new and improved" features with respect to its use as a chassis. The Clean Genome *E. coli* comes with increased electroporation efficiency (see Chapter 2), and introduced plasmids are more likely to remain functional than in commonly used laboratory strains.

These changes may well benefit engineering efforts, and the chassis may be a commercial success. Yet it still contains more than 3,600 genes, and the program by which they run the cell is far from understood. The fact that some of the new features have yet to be explained suggests pursuing a chassis with an even greater reduction in complexity.

Reenter Tom Knight at MIT, who, as part of his program to reconstitute the power of VLSI circuit engineering in BioBrick parts (see Chapter 7), is well aware that the effort could benefit from a chassis with standard interfaces and predictable behavior. His approach is to construct a "minimal genome" by starting with a very simple bacterium and making it even simpler by deleting all the genes not needed for basic metabolism and growth.

Knight's chassis of choice is an innocuous bacterium named *Mesoplasma florum*. It is nonpathogenic to any known organism, contains only a few hundred genes, and grows very quickly in the laboratory. Knight recently collaborated with the genome-sequencing facility at the Broad Institute to completely sequence *M. florum*. Because of the extent of automation now built into the sequencing process, the majority of the task required a mere two hours (see Chapter 6). Knight dryly describes the response to his proposal to sequence an organism with "only" 800,000 bases: "Sure, Tom, we'll sequence that for you—but what do we do with the rest of our coffee break?"[20]

With the completed genome in hand, Knight is proceeding to simplify the instruction set to its minimal content. Knight plans to "redesign the genome of the species from the ground up, with the goal of creating a very well understood, carefully crafted organism suitable as a simple living sub- strate for the nascent engineering discipline of synthetic biology."[21]

This redesign is itself an experiment, because it is not yet clear how even the putative minimal set of genes interacts to support life. It is also unclear how long the project will take or if deleting genes will result in the simple chassis that Knight expects.

Synthetic Genomes

Craig Venter, famous for his contributions to sequencing the human genome, wants to build whole functioning genomes from scratch. In early 2008, a team from the J. Craig Venter Institute published a paper in *Science,* describing the first chemical synthesis and assembly of a full bacte- rial chromosome.[22] The article, with Dan Gibson as first author, reported a method to *synthesize* the ~580,000-base chromosome from *Mycoplasma genitalium* but not the *use* of that chromosome to run a cell.

The goal of these efforts is to enable "synthetic genomics," which team member Clyde Hutchison defines as "chemical synthesis and installation of genomes in cells."[23] There is a somewhat subtle distinction between this goal and synthetic biology as defined by most of its practitioners, which is not necessarily to build whole functioning genomes but rather to construct new synthetic biological circuits to accomplish specific ends, whether for greater understanding or economic impact. Most such systems consist of only a few genes and are thus substantially smaller than the whole syn- thetic genome that Venter and his team are pursuing.

It is well worth understanding a few technical and economic details of the synthetic *M. genitalium* project, because the construction effort served as a stepping stone to a much more powerful technology demonstration later in 2008.

Gibson et al. ordered 101 pieces of DNA, each about six kilobases long, from three different commercial suppliers and then developed a method for parallel assembly of those pieces into the ~580,000-base genome in just a few steps. The commercially supplied fragments were first assembled into twenty-five larger pieces using standard techniques (that the authors called A-series). Two further rounds of assembly resulted in four large, quarter- genome C-series fragments. In the final step, those four quarter-genomes were assembled into a contiguous whole within a yeast host by exploiting that organism's native DNA-handling mechanisms. Those mechanisms are not entirely understood, but they can clearly be exploited by humans to assemble long pieces of DNA. A major advantage of the method is that introducing more biological technology into the process reduces the amount of human handling of large, fragile pieces of DNA.

While the Gibson paper did not directly address the issue of cost, given current prices, one can estimate the team probably spent between $500,000 and $1 million just on DNA synthesis. My estimate for the labor involved in developing the rest of the assembly techniques is in the neighborhood of (at least) five person-years over a couple of calendar years. The full-time equivalent, or FTE, cost for a PhD-level scientist, including salary, benefits, and laboratory costs, is about $250,000 per year in the United States, which means that developing the assembly effort adds another roughly $1.25 million to the cost of the project. This brings the total estimated cost to about $2 million, for just the one experimental assembly.

Clyde Hutchison notes that "[figuring out] the construction step took several years," and after that effort the mechanism by which yeast assembles the synthetic genome remains "a little mysterious."[24] Despite this uncertainty, the success in assembling the quarter-genome fragments into a whole chromosome led the team to wonder whether yeast could do more of the work.

In late 2008, Dan Gibson and his colleagues demonstrated that yeast could be used to assemble the twenty-five A-series pieces into the full *M. genitalium* chromosome in a single step.[25] The assembly required only three days, most of which was spent waiting for the yeast to grow. In their paper, the team wonders whether the genome could be directly assembled in yeast from the original 101 fragments. It is unlikely we will have to wait long to find out. It may be that yeast can be used to build artificial chromosomes as large as several million bases, in the size range of yeast's own chromosomes.

If this speculation is borne out in the laboratory, the result will be a method that enables assembly of chromosomes about the size of Clean Genome *E. coli*. Because the details remain "a little mysterious," there may yet be constraints discovered that limit just which sequences can be successfully assembled. Nonetheless, combined with an existing widespread industry that regularly supplies synthetic DNA fragments of five to ten kilobases in length, rapid assembly in yeast will put the ability to build a wide range of DNA genomes in the hands of scientists, entrepreneurs, and other interested parties worldwide. Craig Venter is certainly thinking along these lines: "What we are doing with the synthetic chromosome is going to be the design process of the future."[26]

The Venter Institute team has already produced a reduced version of *M. genitalium* by experimentally deleting every gene, one at a time, and then discarding genes found to be unnecessary under particular laboratory growth conditions.[27] If the resulting minimal genome were constructed from scratch, coding for an organism that Venter calls *Mycoplasma laboratorium,* it would contain only 382 of the original 517 genes and might serve as another interesting chassis.[28]

The Utility of Genome Synthesis

It isn't clear that synthesizing whole genomes will have immediate economic consequences. Future construction efforts that employ yeast will still require the purchase of many long pieces of synthetic DNA, which will remain fairly expensive for years to come (see Chapter 6). Thus it seems more likely that a reduced genome of some sort—created either through synthesis or through deletion—will be used as a chassis in the near term and grown using standard methods, with synthesis used only to produce customized programming.

Single-step DNA assembly in yeast could be useful here, too, providing a tool to rapidly assemble metabolic pathways from many short DNA sequences. Zengyi Shao and colleagues at the University of Illinois demonstrated precisely this sort of application in late 2008.[29] The researchers assembled functional metabolic pathways from three, five, and eight separated genes in a single-step process nearly identical to that demonstrated at the Venter Institute. If used with libraries of gene variants, where the sequence of every gene in a synthetic pathway might be varied at the same time, assembly in yeast could be used to simultaneously test thousands of different mutations in different genes. This would provide a powerful tool to supplement existing metabolic engineering techniques and potentially to speed identification of useful gene and pathway variants.

But beyond the simple cost of assembling DNA is the cost of labor involved in understanding and building out a new metabolic pathway with economic value. Jay Keasling's work on artemisinin highlights the difficulty of weaving a complex pathway into an existing genome, particularly when the new pathway shares genes and metabolic intermediates with the host. Unless composable parts allow a true separation of the function of the chassis from the circuit programming, the twelve or so proteins demonstrated in the artemisinin pathway may be a practical upper limit on engineering capabilities in the near term.

In the slightly longer term, the more-interesting numbers for synthesis may, therefore, be only ten to fifty genes and ten thousand to fifty thousand bases. Genetic circuits of this size may be the upper limit for systems with economic value for many years to come. More-important questions for would-be genome engineers are, How will the costs fall for constructs of this size? When will DNA of that length be available in days or hours instead of months? How soon before one can buy or build a desktop box that prints synthetic DNA of this length? I don't have specific answers for these questions, but there is clear market pressure to reduce costs and to develop desktop genome-printing capabilities, which I will address in later chapters.

Regardless of the speed of technology development, reductions in capital and labor costs will continue. These trends will support access to, and development of, sophisticated biological technologies in all corners of the world. Exploring the consequences of that proliferation will occupy the remainder of this book.

CHAPTER NINE

The Promise and Peril of Biological Technologies

B IOLOGY IS THE EPITOME of a dual-use technology. All of the tools and techniques that promise progress in basic science and that enable new vaccines can be put to nefarious uses with equal ease.

Our response to infectious organisms such as influenza and the SARS virus are excellent scientific and policy test cases of our readiness for future threats. Only by openly studying pathogens that cause epidemics, publicly discussing the results, and publicly preparing our defense can we hope to be ready for both human creations and natural surprises.

How the Flu Virus Is Put Together

Influenza viruses are difficult and dangerous to work with at the laboratory bench. The ability of these viruses to cause disease in other organisms—their pathogenicity—remains poorly understood. Appreciating the danger and value of the methods used to study infectious viruses requires another cartoon-level exploration of the relevant technology (figure 9.1).

Influenza is an RNA virus; its genes are encoded not on double-stranded DNA, but rather in a genome composed of single-stranded RNA. Complicating matters further, the genome is negative-strand RNA. The genome must be copied into positive-strand RNA—mRNA—before it is translated into protein. Molecular biology as a science is just now producing tools

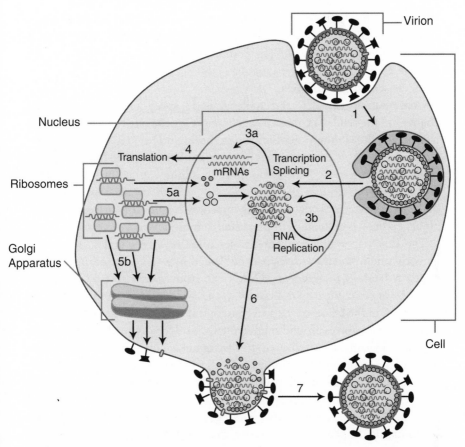

Figure 9.1 Flu virus reproduction. (1) One or more virus particles infect a cell. (2) The viral RNA genome and some original viral proteins enters the nucleus. (3a) The genome is transcribed into messenger RNA (mRNA), and simultaneously (3b) copied. (4) The viral mRNA is translated by the host cell ribosomes into proteins, some of which are (5a) used to produce additional copies of the genome, while others are (5b) further processed by the cell and displayed on the cell membrane. (6) New virus particles form from the copied genome and from host membrane that contains viral proteins. (7) New infectious virus particles are released.

that enable scientists to manipulate and thus investigate negative-strand viruses; it is still very difficult work.

Influenza viruses have just eleven genes distributed on eight chromosomes (sometimes called "gene segments," with more than one gene per segment). Those chromosomes are packaged, along with several kinds of viral proteins, within an envelope composed largely of a membrane stolen from a host cell in which the virion—a single infective viral particle—was produced. Among the proteins carried along by the virus is an RNA polymerase that copies the negative-strand genome into mRNA that host cells then translate into proteins. The complexity of both the viral packaging and the coding strategy have long made doing any molecular biology with influenza extremely arduous.

Contributing further to the challenge is the speed with which the virus changes in the wild. The most frequent ways for the virus to evolve appear to be through (1) mutation of individual bases and (2) "reassortment," the exchange of whole chromosomes made possible when more than one virus infects the same cell at the same time. In addition, the viral RNA polymerase makes frequent copying mistakes, which means that many sequence variants are produced during infection, further enhancing the virus's ability to evolve in the face of pressure from drugs and vaccines.

The first paper demonstrating a so-called reverse-genetics system for constructing influenza viruses was published in 1990.[1] The authors introduced a method to build an RNA virus from constituent parts, which could be of synthetic origin. Because of the difficulties inherent in working with RNA in the laboratory, the technological strategy developed to handle RNA viruses revolves around simplifying the problem by working with DNA instead. Typically, researchers first construct plasmids—that ubiquitous bit of biological technology introduced in Chapter 2—that contain DNA versions of viral genes, which are then reverse transcribed into viral RNA (vRNA) in vitro (figure 9.2).

The authors of the 1990 paper had the goal of eliminating the "difficulty in modifying the genomes of negative-strand RNA viruses that [has slowed our] progress in understanding the replication and the pathogenicity of the negative-strand virus groups." Looking ahead, they also noted that "the ability to create viruses with site-specific mutations will allow the engineering of influenza viruses with defined biological properties."

Improvements in reverse genetics were published throughout the 1990s. A 1999 paper introduced "packaging plasmids" that simplified viral assembly.[2] The specific sequences on the packaging plasmids constituted a program that (1) choreographed the behavior of proteins within the host cell to first make viral proteins necessary for transcribing the DNA into vRNA and (2) then directed the construction of additional proteins that, in effect, act to mimic viral infection by packaging the vRNA into active viral particles. A further advance in 2005 reduced the number of required packaging plasmids from twelve to two, thereby dramatically improving the efficiency of building artificial viruses.[3]

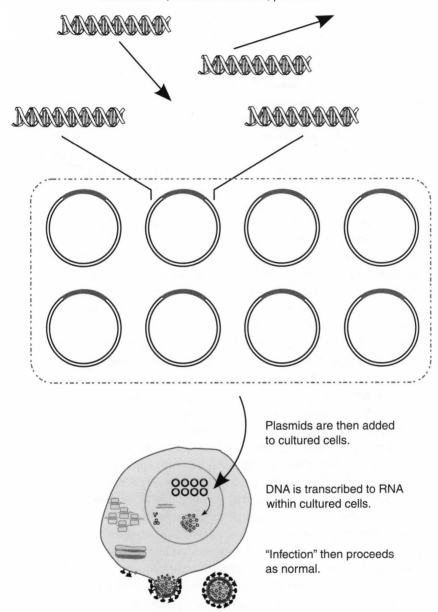

Figure 9.2 Assembling a synthetic flu virus via reverse genetics.

The Consequences of Reincarnating
a Pandemic Virus

In the fall of 2005, several high-profile academic papers described the genomic sequence of the 1918 Spanish flu, responsible for more than 40 million deaths worldwide between 1917 and 1919. This feat was possible because Jefferey Taubenberger at the Armed Forces Institute of Pathology (AFIP) in Rockville, Maryland, determined the sequence of the virus using RNA fragments he recovered from tissue samples stored in government repositories and from the lung of a victim buried in the Alaskan permafrost. The physical sequence was rebuilt by a team led by Terrence Tumpey at the Centers for Disease Control and Prevention (CDC). The flu genome publications were reviewed by the U.S. National Science Advisory Board for Biosecurity (NSABB), which is composed of knowledgeable members of academia and government agencies, and the publications were determined to be in the best interest of the public.

Critics denounced the reconstruction of the sequence as the height of folly, asserting that the project was of questionable scientific and public health benefit and that electronic rather than laboratory study of the sequence information would be sufficient to discover the virus' secrets.[4] Nonetheless, within a year of the flu reconstruction, articles appeared that validated the decision to rebuild the virus. In particular, the experimental results, according to J. C. Kash et al., "indicated a cooperative interaction between the 1918 influenza genes and show that study of the virulence of the 1918 influenza requires the use of the fully reconstructed virus."[5] In early 2007, this result was extended to primates in a study that investigated in monkeys the molecular mechanisms that the host mounts in defense against the virus.[6] That paper described the important role of a vigorous early innate immune response in both controlling the virus and causing tissue damage to the host, which may begin to explain why the Spanish flu killed otherwise healthy young adults at such high rates.[7] Research continues into why the 1918 strain of the flu was so deadly.

At the time the 1918 flu sequence was announced, several high-visibility editorials and op-ed pieces questioned the wisdom of releasing that information into the public domain, suggesting the sequence could be used to create bioweapons. From the perspective of LEGO-style biology, it would appear that all you have to do is plug the appropriate DNA sequence into the packaging plasmids, dump those into a pot of mammalian cells, and wait for infectious viruses to spread throughout the culture. In reality, the process is full of high art and skill, and it is no simple matter to take synthetic DNA and from it create live, infectious negative-strand RNA viruses such as influenza. This is a crucial point. I have discussed this issue with a number of RNA virus experts, including some involved in sequencing and building flu strains, and they universally say reproducing the flu genome is presently quite difficult even for experts.[8] Unfortunately, the scientists involved have engaged in far too little public discussion of the threats posed by reconstituted pathogens.

More importantly, however, the threat from a modern release of the 1918 flu virus is not as dire as many fear. During the 2007–2008 season, the CDC found that 26 percent of samples positive for influenza viruses contained the H1N1 subtype, named for proteins on the outside of the virus and identical to the proteins on the 1918 strain.[9] Therefore, the subtype continues to circulate widely in the population, and most people now have some immunity to it.[10] Without minimizing any illness that might result from release of the original 1918 flu virus, we can conclude that suggestions that any such event would inevitably be as deadly as the first go-round appear to be overstated.

Nonetheless, as in the case of every other biological technology described in this book, it is inevitable that the technology to build RNA viruses will become widespread. Although it is challenging *right now* for even experts to recreate live, pathogenic influenza viruses from synthetic DNA, I have no doubt that over time the relevant skill set will eventually reach individuals with considerably less good sense about how to safely handle the resulting organism. There will probably even be a kit available someday that reduces the expertise to following a recipe, and there may well be an automated platform that implements the recipe with minimal human participation.

The consequent threat to public health and safety from proliferating skills and technology will be substantial. We will eventually require constant vigilance and the ability to detect threats, and to either preempt or remediate them, on short notice. As I argue in the next sections of this chapter, guaranteeing public health and safety requires significant and rapid maturation of technologies that enable biological engineering and, consequently also enable increased threats.

We're stuck; there are no two ways about it. We require new technology to deal with threats, technology that can be developed only within the context of a diverse and capable bioeconomy. Yet the existence of that technology will enable widely distributed use for both beneficial and nefarious purposes. Dealing with emerging biological threats will require better communication and technical ability than we now possess, as directly revealed by progress that resulted from rebuilding and publishing the 1918 influenza genome. Open discussion and research are crucial tools to create a safer world.

To be clear, I do not dismiss the potential of DNA synthesis technologies to be used for nefarious purposes, but rather I have come to the conclusion that it is a looming rather than imminent threat. A recent survey of the difficulty of constructing pathogens de novo concluded that "any synthesis of viruses, even very small or relatively simple viruses, remains relatively difficult."[11] The report then acknowledges that the risk is likely to increase; it is a virtual certainty that someday a synthetic pathogen will emerge as a threat.

We don't have to wait for that day to find out how we might respond to such threats; we are presently at the mercy of many emerging diseases. While the threat of *artificial* pathogens appears minimal at present, there exists a clear and present danger from *naturally occurring* pathogens against which we are demonstrably incapable of defending ourselves. The combination of increased intrusion of humans into previously undisturbed environments and rapid transportation of people and goods around the globe is raising red flags for observers who are concerned about public health and national security. With respect to reconstructing the 1918 flu, Jefferey Taubenberger concludes that "there can be no absolute guarantee of safety. We are aware that all technological advances could be misused. But what we are trying to understand is what happened in nature and how to prevent another pandemic. In this case, nature is the bioterrorist."[12]

It is this problem of naturally occurring pathogens, a problem immediate and pressing, that must define the path of our scientific and technological investment in the near future.

Emerging Infectious Diseases Are Test Cases for Artificial Pathogens

Over the last two centuries the use of vaccines has dramatically altered human and animal susceptibility to disease. While antibiotic drugs are available to treat most bacterial and fungal infections, only very limited treatment options are available for viral infections; vaccination remains the primary countermeasure against viral threats.

Most vaccines consist of a weakened or killed pathogen, or organisms closely related to pathogens that confer cross-strain immunity. Viruses destined for use in vaccines are generally produced in cell culture or grown within fertilized chicken eggs in sterile facilities. While this strategy has proven useful against many pathogens, its future utility is in question. A gradual reduction in production capacity has resulted in recent vaccine shortages during demand spikes.[13] Commercial flu-vaccine production has become unpopular with manufacturers because of the cost of maintaining appropriate facilities coupled with uncertain year-to-year demand, which together make it hard to generate, or at least forecast, a consistent profit. Moreover, even with recently announced significant government participation in the market, it is far from obvious that existing technologies enable production of a sufficient number of doses to protect large populations against a worldwide pandemic.[14]

The resulting "vaccine gap" is more than a weak link in the domestic public-health infrastructures of developed countries. It is a clear example of the threat posed by emerging infectious disease to the interests of all nations. The growing economic interdependence of the Americas, Europe, Asia, and Africa critically increases biological risks globally.

Looking forward, estimates of the worldwide economic consequences of an influenza pandemic are upward of a trillion dollars.[15] A large portion of this figure is due to travel restrictions and to lost trade that would be an inevitable consequence of an outbreak. Limiting global trade has further downstream consequences because of the present reliance of the scientific enterprise on rapid shipping of reagents and laboratory supplies. Disruption of shipping could affect research and countermeasure development by limiting access even to basic material such as gloves, disposable labware, and reagents. As a result, physical and economic security would benefit from disseminating not just vaccines but also vaccine production as broadly as possible. We would be well served by a broad investment in pandemic preparedness.

While there is considerable long-term potential for artificial threats, these remain hypothetical and unlikely in the immediate future.[16] Naturally occurring viral agents provide real examples of rapidly emerging threats. Our responses to the SARS virus and the emergence of the new avian influenza strains provide a context for examining our present technological capabilities.

The Response to SARS

In 2003 a new virus appeared in human populations that caused a rapid and unpleasant death in about 10 percent of those infected. Severe Acute Respiratory Syndrome (SARS) was caused by a coronavirus (CoV) previously confined to animals in southern China. The disease was highly contagious, spreading rapidly across several continents via air travel, with recognition of the event slowed by suppression of relevant information within China. The last human infection that originated from the initial

SARS-CoV outbreak occurred in June of 2003—lab accidents led to several subsequent individual infections—with an official total of 8,096 cases and a final death toll of 774.[17]

Press reports summarizing the SARS episode often hold up the rapid sequencing of the pathogen as a pinnacle of modern technology applied to counter a threat. This achievement is construed as a victory over the virus, as if sequencing helped defeat the epidemic. However, while sequencing is likely to be crucial in combating future emerging infectious disease, technology played only a small role in limiting casualties from SARS.

Whereas the original human case of SARS was forensically traced back to November of 2002, this conclusion was not reached until after the epidemic had subsided and is thus entirely retrospective in nature.[18] The first real-time indication of an emerging viral threat came via Dr. Carlo Urbani's report to the World Health Organization in late February of 2003. Travel advisories were issued over the following four weeks, and the WHO documented the existence of "super-spreaders"—nearly asymptomatic carriers who spread the virus widely. The SARS sequence was announced in the middle of April 2003, but this occurred even before the SARS-CoV was unambiguously identified as the clinical disease agent.[19]

Another six weeks brought the last of the deaths from the natural outbreak. Yet it was not until October that Ralph Baric's group at the University of North Carolina announced a reverse genetics-based system for working with the virus in the laboratory (the protocol was working in the lab as early as June 7).[20] That is, very little molecular biology could be done on the virus—i.e., no basic science and certainly no vaccine development—until well after the epidemic had subsided. Even this belated capability was achieved only because Baric had invested the time and effort several years previously to sort out suitable laboratory techniques. These were first published in 2000, an example of curiosity-driven research playing a crucial role in responding to a subsequent public health crisis.[21]

After the SARS-CoV reverse-genetics system was announced, another six months passed before the first working mammalian vaccine was demonstrated: this was a full year after the height of the epidemic.[22] While some delay in announcing both the reverse-genetics system and the vaccine can be attributed to the academic publication schedule, and while the initial outbreak reporting may have been delayed by poor human-decision-making, it is clear that the technological response severely lagged behind the pace of the virus.

Moreover, epidemiological modeling performed after the epidemic indicates that SARS was successfully contained by public health measures only because of "the moderate transmissibility of the pathogen coupled with its low infectiousness prior to clinical symptoms."[23] In other words, it appears in retrospect that the virus may not have been capable of causing a truly devastating pandemic. If the virus had been characterized by slightly higher transmissibility or a slightly higher infectivity prior to the emergence of symptoms, then public health measures may well have been insufficient to halt disease spread.[24] These speculations only serve to further highlight the technological deficit revealed by the timeline of the SARS epidemic (table 1.1).[25] Fortunately, public health measures *were* sufficient for this particular virus, which teaches us that strengthening the public health apparatus is crucial to preparing for future outbreaks. However, without minimizing the lives saved through public health measures, we must acknowledge that human intervention may have only marginally affected the course of a virus that was less deadly than initially feared. The next time we may not be as fortunate.

Influenza: The Next Virus of Concern?

A virus circulating in a population may have changed substantially by the time a vaccine is ready. The high mutation rate of influenza viruses and long lead times for producing vaccines using current technologies make it difficult to match a vaccine to a specific novel strain.

Geography may also play a role in defeating vaccines. Antibodies produced by ferrets against the original human H5N1-influenza vaccine candidate (isolated in Vietnam) are strongly reactive against influenza isolates from Vietnam but only weakly reactive against isolates from Indonesia and from large parts of China.[26] Conversely, ferret antibodies produced against an Indonesian isolate reacted only weakly against isolates from Vietnam and China. Because of sequence divergence in the wild, antibodies generated in response to the strain from Vietnam are limited in their ability to neutralize the Indonesian and Chinese strains, and vice versa. Another way to say this is that a vaccine generated using one strain may be of only limited use in cross-priming the immune system to recognize a different strain.

Even more troublesome is the recent observation of a genotype in the wild composed of pieces of previously seen viruses.[27] Not only are the avian H5N1 isolates diverging in the wild to the point that they may not cross-prime mammalian immune systems, but they are also actively swapping parts. The viruses are clearly able to exchange innovations on short time scales, thereby producing new potential threats.[28] In response to these developments, in 2006 the WHO advised scientists begin work on H5N1 vaccines based on isolates from Indonesia.[29] Note that this does not imply immediate production of functional vaccines based on these isolates.

The situation is further complicated by the demonstration that flu strains

TABLE 9.1   Timeline of the SARS epidemic

2002	Nov. 16	Forensic identification of atypical pneumonia in Guangdong, China.
2003	Feb. 26	*First case* of unusual pneumonia reported in Hanoi, Vietnam.
	Feb. 28	WHO officer Carlo Urbani examines patient.
	Mar. 10	Urbani reports large outbreak to WHO main office.
	Mar. 11	Outbreak of mysterious respiratory disease reported in Hong Kong.
	Mar. 12	WHO issues global alert.
	Mar. 29	Carlo Urbani dies.
	Apr. 4	SARS added to U.S. list of quarantinable diseases.
	Apr. 9	WHO team reports evidence of "super spreaders".
	Apr. 12	Canadian researchers announce first sequence.
	Apr. 14	CDC announces similar sequence.
	Apr. 16	WHO officially identifies the disease agent.
	May 30	SARS-CoV sequence published in *Science*.
	Jun. 1	*Last deaths* from "natural" outbreaks.
	Oct. 3	Baric group publishes reverse-genetics system with Urbani strain.
2004	Apr. 1	Yang et al. publish DNA vaccine in a mouse model in *Nature*.

can evade a vaccine within a single flu season, as demonstrated through comparisons of annual influenza samples via genome sequencing. After the 2003–2004 annual vaccine was already specified and in production, the most significant population of annual flu strains actually circulating that year adopted a gene from a minor population, thereby producing a new strain "for which the existing vaccine had limited effectiveness."[30]

The clinical ramifications of this event are still not completely clear, with estimates of vaccine effectiveness—that is, successful prevention of symptoms of respiratory illness—ranging between 3 percent and 95 percent.[31] The 2003–2004 wave of infections was associated with an unusually high number of pediatric deaths, and there was an anomalous jump in overall pneumonia- and respiratory-related mortality during the early part of 2004.[32]

In 2008 the annual flu vaccine "failed" again, with one study in the United States showing only a 44 percent effectiveness versus an expected value of 70–90 percent.[33] The Centers for Disease Control and Prevention reported that pneumonia and influenza mortality exceeded epidemic levels for nineteen consecutive weeks in early 2008. These data serve as explicit historical evidence that influenza viruses can escape the protective effects of a vaccine via reassortment in much less time than is presently required to make the vaccine.

These results underscore the following conclusion: because the time scale defining our technological response is so much longer than the time scales on which the threat can arise and change, it would behoove us to invest heavily in new technologies that have fundamentally different characteristics than those employed in the existing production of annual flu vaccines.

Addressing New Threats

New vaccine technologies are required to protect populations threatened by bioterrorism, emerging diseases (SARS) or reemerging infectious diseases (tuberculosis, plague), and pandemic flu strains, all within the context of rapid global travel. The capability of infectious disease to inflict great harm is illustrated well by the fact that disease far outweighed other causes of hospitalization for the U.S. military over the five decades from World War II to the first Persian Gulf conflict.[34]

Any new vaccine technology will have to meet certain design challenges to address the above concerns. Vaccines should ideally be produced rapidly (within weeks of identifying a threat) and inexpensively (to fall within the practical threshold of public health budgets around the world that amount to only a few dollars per capita annually). The global flow of people, live animals, animal products, and general biological material adds to the likelihood that disease will be transmitted before surveillance identifies a threat. It is only by fortunate circumstance that in each of two recent outbreaks, SARS and the West Nile virus, multiple patients were seen by a single clinician, thereby allowing early identification and resulting in a reduction in the spread of those diseases.[35] Although quarantine was employed in both cases, isolation is generally not considered an effective means of containing infectious disease, in part because a small lapse in containment can result in fresh and widespread outbreaks.[36] This is particularly a problem with pathogens that spread before symptoms become apparent.

Of crucial current relevance, pandemic flu viruses have historically spread widely before being detected, which is one reason why a vast number of doses of vaccine and antiviral drugs will be necessary to combat an outbreak. Analyses of recent human outbreaks of H5N1 in Turkey and Indonesia concluded that patients were infected an average of five days prior to the appearance of symptoms, outside the window recognized for effective use of antiviral drugs.[37]

Our general problem, highlighted by both SARS and Avian influenza, is that our present technological base is not sufficient to deal with new threats. Humans have not previously succeeded in controlling a *rapidly spreading, novel* pathogen with either drugs or vaccines. Thus any such future action is, in effect, an example of an attempt to engineer a complex system with even fewer tools, and less real-time information, than exist for synthetic genetic circuits. At this point we must rely on mathematical models to gain at least some insight into the potential consequences of various public health and vaccination strategies.

Those models suggest that, for the time being, our best strategy is to push technology forward as rapidly as possible in order to deliver solutions that can used to rapidly and broadly counter disease outbreaks. Synthetic vaccines are potentially one such solution.

The Continuum of Vaccines: From Natural to Synthetic

There is neither an obvious nor a discrete boundary between (1) "natural" vaccines, which are composed of organisms purified from patients and then grown to quantity in eggs or cell culture, and (2) "synthetic" vaccines, whose genomes are modified or that contain little or no genetic material from the original pathogen. Many vaccine strains are produced by breeding microbes to be less pathogenic. Newly developed technology further blurs the distinction between natural and synthetic.

The advent of rapid whole-genome sequencing allows identification and isolation of gene sequences from pathogens. Emerging vaccine technologies will enable using only those genes that generate an immune response and leaving out the genes that pathogens use for reproduction or that cause actual disease. In the lab, selected genes may be used to directly produce proteins for use as antigens, included in less-pathogenic organisms that serve as vaccine vectors, or inserted into DNA plasmids for epidermal injection (see table 9.2). All of the these technologies have proved useful in selected circumstances, such as protecting monkeys, horses, fish, chickens, ferrets, and cats from the Ebola virus, the Marburg virus, and various flu and encephalitic viruses. But, as of early 2008, none are presently approved for use in humans.

TABLE 9.2   Types of vaccines

Natural vaccines	
Inactivated viruses	Pathogens inactivated by UV irradiation, photodamage, or chemical modification. (Example: influenza)
Attenuated active viruses	Live viruses that provoke a protective immune response but that are not themselves pathogenic. Most often produced by "passaging" the virus in other host species for many generations, which results in variants with large numbers of mutations that reduce virulence. (Examples: MMR, adenovirus)
Synthetic vaccines	
Subunit vaccines	Recombinant proteins produced from pathogen genes inserted into expression vectors.
Virus-like particles (VLPs)	Self-assembled viral particles that contain pathogen proteins but no DNA and that are unable to replicate.
Microbial vectors	Harmless viruses or bacteria carrying genes that code for pathogen coat proteins, which elicit a protective immune response against the whole pathogen.
DNA vaccines, also known as gene vaccines	Naked DNA plasmids, or plasmids packaged in viruses, containing genes that code for pathogen proteins.

Synthetic vaccines are possible only through the application of new technology. Working with RNA viruses, such as influenza, requires relatively sophisticated molecular techniques in order to produce variant versions of the genome. Yet if produced via current technology, a new vaccine strain must still be grown either in chicken eggs or in mammalian cell culture. These techniques introduce a time lag for turning a new pathogen into a vaccine. The most useful functional delineation of vaccines is based on the type of threat against which they are useful as countermeasures, in particular the immediacy of the threat. The time period required for production and distribution thereby defines the utility of a vaccine when a threat is imminent.

Time Scales and Rapid Response

Recent epidemiological models suggest that mass vaccination programs will significantly reduce the total number of infected persons only if they begin within 90 days of an influenza outbreak, with morbidity and mortality reduced only marginally when vaccinations begin after day 120.[38] Thus the capability to act quickly is critical from a biosecurity perspective.

The U.S. Department of Health and Human Services has recently awarded more than a billion dollars to firms working to produce vaccines in cell culture, with the expectation of both more-rapid and higher-volume production. But ramping up production capacity will take many years, in part because that capacity must ultimately be supported by the very competitive market for annual flu vaccines, for which in most years there now exists a "huge oversupply" in the United States.[39] While cell culture does diversify the technology base for producing annual vaccines, it does not necessarily improve the speed with which they can be developed.

New live or attenuated vaccine strains have historically been treated as entirely novel drug formulations, requiring a full course of safety and efficacy testing, thereby considerably lengthening the time between disease identification and availability of a vaccine. Even with relaxed regulatory requirements during a pandemic, the typical time scale quoted by academic experts and government officials for fielding a new flu vaccine is six months. However, egg-based or cell culture-based production of entirely new vaccine strains may realistically require as long as nineteen months.[40] That is, when we have an indication that a particular strain of a pathogen may of concern, we may have time to include it in the vaccine mix for use one to two years in the future. In the case of a surprise, the present vaccine production infrastructure may well be too slow to be of much assistance. Fortunately, the tools already exist to quickly build synthetic vaccines.

In the event of a flu pandemic, the very same tools used to reconstruct the 1918 influenza virus can be used to rapidly synthesize a vaccine. The gene-synthesis infrastructure decried as a security risk by many critics can be used to construct genes for use in DNA vaccines. Similar techniques can be used to select genetic material from other pathogens for use in many different strategies to produce vaccines. However, this technology is not yet perfected. DNA vaccines are a true first-generation synthetic biology project; not only are the vaccines themselves constructed via rearrangement of existing genes and synthesis of new ones, but the rules governing immune response to such vaccines are not sufficiently understood to enable deterministic design. All of which will improve dramatically in the coming decade.

Use of viral vectors to deliver a DNA vaccine for H5N1 was recently demonstrated in mice and poultry, with the vaccine produced *less than four weeks* after pathogen sequence identification.[41] We can expect results from human clinical trials within the next few years, and the general strategy is well matched to prompt distribution of countermeasures to rapidly emerging natural or artificial pathogens.

Briefly, a distributed DNA-vaccine production strategy set in the near future might run as follows: New natural or artificial pathogens are first isolated and sequenced using contemporary technology and infrastructure.[42] In the case of new viruses or particularly deadly reemergent strains, some basic molecular biology may be necessary to figure out which viral genes *not* to use in a vaccine. Ebola, Marburg, and influenza viruses, for example, all contain sequences that, when expressed, act to suppress human antiviral responses. Once genes appropriate for inclusion in a vaccine are identified, the sequences would be made available to synthesis facilities and transmitted electronically worldwide. Those facilities could then locally synthesize vaccine-vector plasmids containing antigen genes, followed by distribution to the local population.

Rather than waiting to grow pathogens themselves for use as vaccines, we must accelerate development of technologies that provide for rapidly turning sequence information directly into widely distributed vaccines.[43] More generally, we do not now have the capability to rapidly detect threats, to rapidly understand the biology of new organisms, or to rapidly use any such information in the manufacture and distribution of countermeasures. Finally, returning to the possibility of artificial pathogens, we must also prepare for the day when human hands will be capable of producing threats as dangerous as those due to nature.

The Proliferation of Skills
and Materials Is Inevitable

When it was founded in 1990, the sequencing facility at the Whitehead Institute Center for Genome Research (now the Broad Institute) employed primarily scientists with doctorates to process DNA. Over the years these PhDs were gradually replaced by staff with master's degrees, then bachelor's and associate's degrees. Now many of the staff responsible for daily sequencing operations have completed only a six-month qualification course at a local community college.[44] These technicians are trained in all the steps necessary to shepherd DNA from incoming sample to outgoing sequence information, including generating bacteria that contain DNA from other organisms.

Despite the impressive results described in previous chapters, most students and novice gene-hackers will be much better at conceptualizing and modeling new genetic circuits than actually building them. Where design expertise exceeds practical experience, commercially available kits include simple recipes that allow moving genes between organisms. The process might be slightly more complicated than baking cookies, but it is for the most part less complicated than making wine or beer. This broad distribution of biological technology naturally leads to questions of how it will be applied.

The debate over who will be permitted access to which biological technologies is among the most important of the twenty-first century. It is unlikely that regulation of skills will produce an increase in public safety. The industrial demand alone for skilled biotechnology workers in the United States increased 14–17 percent per year during the 1990s, and many of these workers came from overseas.[45] Not all these workers will remain in the United States, and it is safe to say many of those who leave will make use of their skills elsewhere. If the U.S. government acts to try to limit the practice of certain methods, it will be unrealistic to try to centrally monitor every skilled person in the country. The same is likely to be true in any other country. We certainly cannot "unteach" the relevant skills in order to prevent unauthorized use, and any action to limit the proliferation of skills would cripple that portion of the economy reliant upon biological technologies. Thus, as the technology's total contribution to the economy grows, any restriction on training in this area could have severe negative economic consequences.

Perhaps more problematic than distributed skills will be ubiquitous materials. Effective regulation relies on effective enforcement, which in turn requires effective detection. The extent of illegal-drug production in the United States and previous failures to detect illicit biological-weapons production gives some indication of the challenges of detection and enforcement in the context of widely distributed technology.

The widespread illicit distillation of alcohol during Prohibition in the United States and the proliferation of modern illegal-drug-synthesis labs both illustrate the principle that outlawing chemical products merely leads to black markets, which are more difficult to observe and to regulate than open markets. The knowledge of chemistry required to produce drugs is not significantly less complex than what is required to genetically manipulate organisms. Therefore, experience in regulating the production and use of illicit drugs serves as a useful reference point for any similar efforts to regulate biological technologies.

The practice of synthesizing prohibited compounds is widespread. Approximately eight thousand clandestine methamphetamine (meth) laboratories were seized in the United States in 2001, with more than 95 percent of those being mom-and-pop operations producing less than five kilograms per day.[46] Due primarily to domestic restrictions on precursor chemicals and vigorous law enforcement efforts, the number of methamphetamine labs in the United States has dropped considerably since 2000 and meth-related sentencing has climbed dramatically.[47] The amount of meth seized by the Drug Enforcement Agency (DEA) has fluctuated around 1,700 kilograms since 2000.[48] The total number of methamphetamine-associated "clandestine laboratory incidents" declined sharply in 2006.[49] However, the consequent reduction in supply by mom-and-pop producers has been more than offset by a combination of increased centralized domestic production and imports, both of which appear to be in the control of large criminal organizations.[50] U.S. meth consumption continues to increase, supplied from abroad; between 2002 and 2004, the amount of meth seized at the U.S.-Mexico border increased more than 75 percent.[51]

According to the 2007 National Drug Threat Assessment from the National Drug Intelligence Center at the DEA, the increase in foreign production appears to be concentrating production and distribution technology and infrastructure within drug-trafficking organizations (DTOs): "Marked success in decreasing domestic methamphetamine production through law enforcement pressure and strong precursor chemical sales restrictions has enabled Mexican DTOs to rapidly expand their control over methamphetamine distribution."[52]

Regulation is therefore causing a shift from distributed, domestic production to foreign centralized, criminal organizations. The resulting concentration of capability and capital has enabled DTOs to embark on sophisticated independent technology-development programs that innovate in the areas of drug production, financing, and distribution. To evade increased aerial and marine patrols, drug smugglers have of late been using home-built submarines along the coasts of Australia, Europe, Central America, and the United States. One such vessel, discovered before completion, would have been capable of hauling two hundred metric tons of cargo in a single trip and was described as being built using advanced technology and very high-quality workmanship.[53] As of spring 2008, drug enforcement agents report spotting ten such vessels on average every month, only 10 percent of which are successfully intercepted—it is unclear what fraction of the total in operation are even spotted.[54]

Centralized production and the greater flow of drugs across U.S. borders is consistent with the argument that policing internal to the United States encourages the creation of elaborate and evidently very effective black-market drug-manufacturing and drug-distribution networks. Thus increased enforcement efforts are paradoxically producing an infrastructure that is, according to the DEA, "more difficult for local law enforcement agencies to identify, investigate, and dismantle because [it is] typically much more organized and experienced than [that of] local independent producers and distributors."[55] Beyond the simple physical difficulty of enforcing laws against drug production and distribution, there is a strong economic incentive for producers to evade such regulation.

Methamphetamines cause obvious and extensive medical problems for users, while associated social ills and criminal activity result in substantial human and economic cost. There does not appear to be any easy fix. The underlying problem is that some people are willing to pay for, and others to produce, an illegal and dangerous product of a distributed technology despite harsh consequences for both users and producers.

The False Promise of Regulation

Where there is a market, there will always be attempts to supply it, even when the product is both legally and culturally frowned upon. When the product in question is of sufficiently high value, either monetarily or politically, considerable sums will be spent to produce and supply it. This is as true for biological technologies as it is for illicit drugs, and as true internationally as within any nation.

Numerous examples of intelligence and enforcement failures provide explicit challenges to the notion that the risks posed by mistakes or mischief resulting from biological technologies can be mitigated through regulation. More relevant to threats from biological technologies, a wide array of Western intelligence services failed to uncover the existence of extensive bioweapons programs in Iraq before the First Gulf War.[56] Describing the prelude to the Second Gulf War, Richard Spertzel, a veteran bioweapons inspector with many years of experience in Iraq, stated that he "heard from a reputable source that Iraq had moved most of its [bioweapons] program to the Bekka Valley of Lebanon early [in 2003]."[57] In the larger scheme of things, particularly in the context of the debate about the accuracy of intelligence concerning Iraq's weapons programs prior to 2003, the validity of this claim is neither interesting nor relevant to my argument.

Whether or not that bioweapons program existed, or was moved out of Iraq before the invasion of 2003, an experienced weapons inspector found such claims credible enough to repeat on the record. That is, technology and skills are so widespread that it is easy for an expert to believe production facilities for biological weapons are already beyond the considerable reach of the international intelligence community. In a world in which it is this easy to hide production facilities, how are we to prevent black markets for illicit substances? How are we to limit access to the technologies used to produce them?

Given the potential power of biological technologies, it is worth considering whether open markets are more or less desirable than the inevitable black markets that would emerge with regulation. Those black markets would be, by definition, beyond regulation. More important, in this case, they would be opaque.

The real threat from distributed biological technologies lies neither in their development nor in their use per se but rather in the fact that biological systems may be the subject of accidental or intentional modification without the knowledge of those who might be harmed. Because this may include significant human, animal, or plant populations, it behooves us to maximize our knowledge about what sort of experimentation is taking place around the world.

Some observers view as an immediate threat the proliferation of technologies useful in manipulating biological systems: passionate arguments are being made that research should be slowed and that some research should be avoided altogether. "Letting the genie out of the bottle" is a ubiquitous concern, one that has been loudly voiced in other fields over the years and that is meant to set off alarm bells about biological research.

If strict restriction of access to technology held promise of real protection, it would be worth considering. But such regulation is inherently leaky, and it is more often a form of management than blanket prohibition. Certainly no category of crime has ever been *eliminated* through prohibition. More important than any tenuous safety resulting from regulation, however, is the potential danger of enforced ignorance. We must decide not whether we are willing to risk damage caused by biological technology but whether limiting the general direction of biological research in the coming years will enable us to deal with the outcome of mischief or mistake. We must decide if we are willing to take the risk of being unprepared.

Restricting research will merely leave us less prepared for the inevitable emergence of new natural and artificial biological threats. Moreover, it is naive to think we can successfully limit access to existing pertinent information within our current economic and political framework.

While the most advanced research and instrumentation developments may occur first in fully industrialized countries such as the United States, where export might be controlled, other countries are developing a skill base that will enable broad domestic utilization of biological technologies. The expansion of iGEM participation is, by itself, an extraordinary measure of the appetite for new biological technologies around the world (Chapter 7). And it is no coincidence that China fielded two of the top-six iGEM teams in 2007.

China has an aggressive program in plant biotechnology, which increased in annual funding from $8 million to $50 million between 1986 and 1999.[58] In September of 2008, a new plan was announced to spend at least another $3 billion by 2020, following this announcement by Premier Wen Jiabao to the Chinese Academy of Sciences: "To solve the food problem, we have to rely on big science and technology measures, rely on biotechnology, rely on GM [genetic modification]."[59] This energetic investment also exists in the Chinese private sector, rewarded by a 30 percent annual growth rate in the biopharmaceutical sector between 2000 and 2005.[60] The national scientific establishment is attempting to lure back foreign-trained Chinese scientists by offering lucrative financial packages.[61] The first governmentally approved gene therapy is available not in the United States nor Japan nor anywhere in Europe but rather in China.[62] A prior premier stated in public that the government would use all means available to it to improve the health of the population, including genetic modification of its citizens.[63] Many other countries are in the process of bootstrapping their capabilities to develop and use biological technologies.

India is tripling funding to its national biotech center, has the aim of spending one-third of its research budget on biotechnology, and is promoting the development and use of genetically modified crops throughout Asia.[64] Singapore has for many years made a practice of recruiting foreign scientists.[65] Taiwan is investing large amounts in biotechnology and is seeking citizens to return home to build up biotechnology in academia and industry.[66]

Given these developments, within the context of the increase in individual capabilities and the independent reduction in cost, it is unrealistic to think biological technologies can be isolated within the borders of any given country. Even if such a regime were implemented, it would merely include those countries that already have a particular technology. We can do little to take technology away from those in whose hands it was developed and resides. The best strategy going forward is in fact to encourage such efforts at all levels in an open environment.

In light of the proliferation of gene-synthesis technology and skills described in Chapter 6, it is clear that any attempt at regulation would have to be implemented within an international framework. A small group of academic and industrial participants active in developing and commercializing synthesis technology is attempting to be proactive on this front. The initial public statement from the International Consortium for Polynucleotide Synthesis suggests:

> As part of the process of improving DNA-synthesis technology, it is imperative that DNA-synthesis firms develop and implement effective biological safety and security procedures, while retaining the ability to deliver high-quality products at low cost and with very rapid delivery times. The full constructive potential of DNA-synthesis technology will be realized only if a governance framework is developed that is compatible with the needs of industry and customers, and that supports best practice in biological safety and security, including the effective deterrence and investigation of any criminal uses of synthetic DNA.
>
> A governance framework that stymies the open commercial development of synthesis technology will retard research and make the challenge of responsibly developing the technology more difficult. Likewise, a regulatory framework that hampers a single country's or group of countries' commercial market without international consensus will drive consumers to the most facile and cheapest available source, and have a limited impact on enhancing global security.[67]

These recommendations are informed by the desire for safety and security, the rapid pace of technological innovation, and the wide global distribution of DNA-synthesis technologies and gene-assembly methods. In such a context, implementing regulation of technologies that are already broadly distributed inevitably drives innovation by users of those technologies, as the example of illicit-drug production demonstrates.

Regulation of work in the lab will be just as problematic as regulation of compounds. Provisions of the USA Patriot Act were intended to improve biosecurity in laboratories working with certain organisms, by instituting both physical security measures and background checks on scientists, with the possibility of restricting access based on nationality. Among the first effects of its passage was a reduction the number researchers in the country working to understand disease-causing pathogens. In an interview in the *New York Times,* Nobel Laureate Robert Richardson lamented the effect of the legislation on research at his home institution, Cornell University, which immediately lost thirty-six of thirty-eight talented researchers working on deadly microbes: "We've got a lot less people working on interventions to vaccinate the public against smallpox, West Nile virus, anthrax and any of 30 other scourges."[68] It is hard to see how this improves the chances of dealing with emerging or intentionally released pathogens. In fact, the anecdote suggests increased regulation will further impair security.

If regulation is not merely an ineffective option but will actually be an impediment to security, how can we attempt to mitigate coming risks? The goal is clearly both to counter mistakes in the laboratory and weapons created from biological components and, ideally, to make such threats irrelevant before they become a problem.

It may be many decades before our understanding of biology provides for the requisite rapid detection, analysis, and response. Fortunately, it is also probably true that we have some time to prepare before both technology and skills become truly pervasive. In the meantime, all parties can work together to lay the groundwork for an increase in security through dramatically improved communication and focused technology development. We must also weigh very carefully the impacts of whatever regulations are implemented in the near term.

CHAPTER TEN

The Sources of Innovation and the Effects of Existing and Proposed Regulations

How do we maximize the pace of technology development while improving physical and economic security? Creating and commercializing new tools and methods is a complicated process. There is an enormous difference between demonstrating functionality in the laboratory and producing a product that people want to use in the real world. As in the case of technologies discussed in prior chapters, government policy and the availability of funding will play important roles in moving biology from the lab into the economy. As we consider how best to foster the development of new technology, it helps to explore where that innovation arises.

What Is Innovation, and from Where Does It Come?

Invention is not the same thing as innovation. Invention has been described as the act of capturing a natural phenomenon for a use by human hands, or of building an object that performs a new task.[1] Innovation may subsume invention but more generally describes the entire process of turning an idea into an object useful beyond the laboratory. The distinction is usually glossed over in the economic literature, in favor of simply studying "innovation." This is in large part because classical economics attempts to understand markets based solely on price. But markets possess no intrinsic means to place a value on invention, let alone a mechanism to distinguish between the value of invention and the value of innovation. Yet both activities are necessary for new technologies, processes, and products to find their way from the laboratory to the consumer.

Theoretical discussions of the connection between invention and innovation in the economy go back at least to Joseph Schumpeter in the 1920s and 1930s. More recently, New York University economist William Baumol has pushed the discussion forward through an argument that "private innovative activity has been divided by market forces between small firms and large, with each tending to specialize in a different part of the task":

> Even though the preponderance of private expenditure on research and development (R&D) is provided by the giant business enterprises, a critical share of the innovative breakthroughs of recent centuries has been contributed by firms of very modest size. These radical inventions then have been sold, leased or otherwise put into the hands of the giant companies, which have then proceeded to develop them—adding capacity, reliability, user friendliness and marketability more generally—to turn them into the novel consumer products that have transformed the way Americans live.[2]

Baumol notes that "the menu of innovations [due to small firms] literally spans the range from A to Z"; several of those innovations are listed in table 10.1.[3]

An Intrinsic Division of Labor

While small firms appear to be an important source of inventions and new technologies, it is large firms that are the major source of actual products. According to the National Science Board, just 167 large companies (with twenty-five thousand or more employees) accounted for 46 percent of total U.S. industrial R&D spending in the year 1997; 1,733 companies with more than one thousand employees contributed another 34 percent. The proportion of R&D spending by large companies remained essentially unchanged up through at least 2003.[4] That a few very large firms spend nearly half the U.S. R&D budget is not a sign that they are poor innovators but rather that the innovation required to get products out the door into customers' hands is extremely expensive.

The computing industry provides an excellent example of where R&D investment comes from and how it gets spent. Large corporations spend large amounts of money getting products to work, and the U.S. govern-

TABLE 10.1   Important innovations by U.S. small firms, 1900–2000

Air conditioning	Heat sensor	Prestressed concrete
Air passenger service	Helicopter	Prefabricated housing
Airplane	High-resolution CAT	Pressure-sensitive Tape
Articulated tractor	scanner	Programmable
chassis	High-resolution digital	computer
Cellophane artificial	X-ray	Quick-frozen food
skin	High-resolution X-ray	Reading machine
Assembly line	Human growth	Rotary oil-drilling bit
Audio tape recorder	hormone	Safety razor
Bakelite	Hydraulic brake	Six-axis robot arm
Biomagnetic imaging	Integrated circuit	Soft contact lens
Biosynthetic insulin	Kidney stone laser	Solid-fuel rocket engine
Catalytic petroleum	Large computer	Stereoscopic map
cracking	Link trainer	scanner
Computerized blood	Microprocessor	Strain gauge
pressure controller	Microscope	Strobe Lights
Continuous casting	Nuclear Magnetic	Supercomputer
Cotton picker	Resonance Scanner	Two-armed mobile
Defibrillator	Optical scanner	robot
DNA fingerprinting	Oral contraceptives	Vacuum tube
Double knit fabric	Outboard engine	Variable-output
Electronic spreadsheet	Overnight national	transformer
Free-wing aircraft	delivery	Vascular lesion laser
FM radio	Pacemaker	Xerography
Front-end loader	Personal computer	X-ray telescope
Geodesic dome	Photo typesetting	Zipper
Gyrocompass	Polaroid camera	
Heat valve	Portable computer	

ment supports a significant amount of basic research in both academic and corporate environments. According to the 1999 National Research Council report *Funding a Revolution,*

> Very little R&D performed in industry [as a whole] is research; most, in fact, counts as development. Even applied research accounts for only about 10 to 15 percent of industrial R&D in computing. . . . Basic research in industry is only about 2 percent of total R&D. When one excludes development from

consideration, government support represented about 40 percent of all computer research, and half of that was basic research.[5]

To summarize the argument: product development accounts for most of the R&D spending of large companies. That is, "R&D" spending is mostly "D," and, as the *Economist* notes, "for some time, the computer industry has, in effect, relied for much of its research on small firms."[6]

The historical record of invention by small firms, in light of the larger share of R&D spending by bigger organizations, requires further explanation. Based on empirical observations, Baumol argues that the division of labor between small and large firms is deeply related to the structure of our economic system. Invention and innovation tend to be implemented well by different kinds of organizations. Most important, the development of fundamentally new technologies in a market economy *requires and explicitly depends upon* the participation of small firms and entrepreneurs: "Small entrepreneurial firms have come close to monopolizing the portion of R&D activity that is engaged in the search for revolutionary breakthroughs."[7]

But why, in a market economy, in which competition supposedly sets the price for any product or service, are entrepreneurs able to outcompete large firms and produce such a disproportionate share of radical innovations? One answer, often given in this era of stock options and (multi) billion-dollar initial public offerings, is that entrepreneurs are looking to win the lottery through hard work that they enjoy. In addition to the "superstar market reward structure" for entrepreneurs who do make it big, Baumol suggests that because entrepreneurs are willing to accept financial underpayment for the "great pleasure" of achieving success, they are thereby "richly rewarded overall." This has explicit consequences for the value of their labor in the market, explaining both the structural mechanism maintaining the difference between Davids and Goliaths and the reliance by the Goliaths on the Davids: "The independent innovative entrepreneur will tend to be the economical supplier of breakthrough innovation to the economy."[8]

This division of labor appears to be a general feature of industries that rely heavily on new technology. A 2006 report prepared for the Small Business Administration examined 192 public and private firms in diverse industries across fourteen years, concluding,

> Industries that are becoming increasingly more technical, as represented by an increase in the employment counts of scientists and engineers, are associated with increasing counts of fast growing new private firms, and negatively associated with counts of fast growing established public companies. Further, we find that an increase in the emphasis on production within industries is nega-

tively associated with counts of fast growing new private companies and positively associated with counts of fast growing established public companies.[9]

Baumol suggests there are deep structural reasons that big firms do not pursue risky innovation and instead invest in gradual innovation with predictable, though incremental, implications for increasing revenues. Large firms tend to be engaged in large markets, generally full of competition, in which every firm must find innovative ways to maintain its customer base. Large firms must also spend additional funds to attempt to capture the next customer, who, because of competition, is usually more expensive than the previous customer. This ongoing struggle gives rise to the notion of "pauper oligopolies," in which large companies may have substantial income but must spend most of that income to maintain market share. Baumol notes that this battle has predictable consequences: "The history of arms races confirms that they can be expected to impoverish the participants."[10] There are always firms that appear to escape the struggle while producing enormous profits—present examples include Google, Apple, and Exxon, which are successful in different markets for different reasons—but the vast majority of large commercial enterprises appear to experience rather less monetary success and rather more struggle for survival.

If Baumol's arguments are generally true, and a division of labor historically characterizes the structure of technological innovation within our economy, there is no reason to expect a different structure to arise in the future development of biological technologies. Moreover, if the computing, automotive, and aerospace industries are any indication of the future course of biological technologies, then existing large companies will survive only through continued access to inventions and early stage products from small companies and start-ups.

Baumol presents the hypothesis that "it is the market mechanism that assigns each type of firm to its differentiated job [and] that assigns the search for radical inventions to the small enterprises and their subsequent development to the large."[11] That is, the division of labor in innovation is a deeply embedded feature of a market economy. Looking ahead to the future role of biological technologies as an industry in the economy, we should expect the same division of labor to emerge between large and small firms. The development of this ecosystem of innovation requires that small firms have access to services, skills, and raw materials and also have a general ability to participate in the market. Consequently, if small enterprises and entrepreneurs developing biological technologies face restricted access to infrastructure and markets, then we should expect less innovation overall.

Rephrasing these observations, it is by no means clear that the industries reliant upon biology will develop into a sector of the larger economy organized radically differently than any other sector of the economy. This suggests that in order to achieve the benefits of biological technologies, we must not just tolerate but indeed *foster* a thriving ecosystem of innovators of different sizes populating different niches. If this assertion is true, we will become more secure as skills and knowledge spread.

In the context of potential regulation, those seeking to limit access to foundational engineering technologies and skills must explain how such regulation will not hinder the formation of an innovative ecosystem, thereby hampering both physical and economic security. In order to maintain progress while improving safety, we must examine very carefully the effects of regulation on the ability to innovate.

Fanning the Fire for Regulation

In June of 2006, the *Guardian* newspaper published a pair of articles announcing that its science correspondent, James Randerson, had purchased "a short sequence of smallpox DNA" and had it delivered to a residential London address.[12] In order to prevent the order from falling afoul of the United Kingdom's Anti-terrorism, Crime and Security Act 2001, Randerson introduced mutations into the sequence to produce three stop codons intended to guarantee safety. When confronted with this fait accompli by a reporter pushing the threat to public safety as a front-page story, Phil Willis, member of Parliament and chairman of the parliamentary Science and Technology Committee, responded as one might expect: "This is the most disturbing story I have heard for some time. There is clearly a massive loophole which needs to be dealt with by regulation or legislation."[13] In the context of an advertised threat of synthetic pathogens, a predilection to regulate might not be a surprising initial response to gene-synthesis technology. The average reader will probably feel some sympathy for Willis's response. Yet there was substantially less to this story than was presented by the reporter.

To be completely clear, what Randerson demonstrated was the use of the Internet, a credit card, and express delivery. He did not demonstrate any skills necessary to *use* the short, seventy-eight-base-pair DNA sequence obtained in the exercise to build a larger construct. Therefore, while the alarm raised by Randerson may have been intended to generate public discussion about a powerful new technology, the DNA order itself was unrelated as a practical matter, save its role in generating publicity.

That publicity served the public poorly. Like the critics of the publication of the 1918 flu sequence, Randerson promulgated the falsehood that producing a live, infectious pathogen is an easy matter: "To build a virus from scratch, a terrorist would simply order consecutive lengths of DNA along the sequence and glue them together in the correct order."[14] As described in the previous chapter, even experts are still challenged by the task of assembling a working genome from short oligonucleotides. The assertion—in the present tense, no less—that "a terrorist would simply" be able to create a pathogen via genome assembly was, at best, misleading.

Randerson encountered substantial criticism for his reporting from scientists and engineers on both technical and ethical grounds. As part of the online discussion of the *Guardian* articles at *Nature.com,* Randerson remained unapologetic and maintained that the issue of unregulated access to DNA synthesis services deserved attention.[15] This is indeed a critical area of discussion for practitioners, policy makers, and the public. But Randerson's focus was far too narrow: "[Regulations] would have to balance . . . scientific progress—which is of course what will protect us against diseases and the actions of a putative bioterrorist—and the need to prevent technology being used for malign ends." The question left unaddressed by Randerson's reporting—the question that logically comes after "Who should be allowed access to DNA synthesis technology?"—is "Will trying to regulate those activities actually improve safety and security?" What if there is reason to believe that implementing regulation may *degrade* safety and security? This potential outcome must be considered before implementing policy.

Examples of Existing Regulations and Recommendations

It might appear obvious that we should outlaw the possession of particular pathogens or toxins and that we should prevent the synthesis of disease-causing organisms. However, defining just which actions should be restricted, and how to accomplish that, while not unintentionally restricting research is no small challenge. What follows is a short survey of relevant policies put in place by the U.S. government.

The Establishment of the NSABB

In order to better evaluate policy and to make recommendations to improve biosecurity, the U.S. government has established the National Science Advisory Board for Biosecurity (NSABB). It consists of representatives from a wide variety of governmental agencies as well as nongovernmental experts, whose official charter is to "provide advice, guidance, and leadership regarding biosecurity oversight of dual use research, defined as biological research with legitimate scientific purpose that may be misused to pose a biologic threat to public health and/or national security." The purview of the NSABB is limited to all "federally conducted or supported" research. Its charge includes developing policy recommendations, as well as reviewing specific research and potential publications, for government-funded work at academic institutions and private corporations. Note that the charter of the NSABB includes no authority over privately funded research.[16]

The National Select Agents Registry

The U.S. government has also taken more direct legislative action to institute biosecurity measures. The Public Health Security and Bioterrorism Preparedness Response Act "requires that all persons possessing biological agents or toxins deemed a threat to public health to notify the Secretary, Department of Health and Human Services (HHS)."[17] The law also establishes penalties for those failing to notify the proper authorities about possession of select agents. In summary, the National Select Agents Registry (SAR) and the law providing for it are an attempt to track and control possession of specific organisms and molecules deemed to pose health and economic threats.

Proscribing Synthesis of Smallpox

In addition to prohibiting possession and transport of materiel, which is fairly straightforward to interpret legally, the United States has already attempted to restrict use of synthesis technology to produce one specific pathogen. The Intelligence Reform and Terrorism Prevention Act of 2004 contained an amendment, inserted at the last minute, that "imposes severe penalties for attempts to engineer or synthesize the smallpox virus. The amendment defines smallpox virus as any virus that contains more than 85 percent of the gene sequence of variola major or variola minor."[18] The amendment went mostly unnoticed for several months.[19] Even when it became widely discussed, virologists examining the bill's small print "[could not] agree on what exactly it outlaws."[20] The broad definition of similarity to smallpox is particularly problematic because, "many poxviruses, including a vaccine strain called vaccinia, have genomes more than 85 percent identical to variola major."[21]

The NSABB, as part of its duties, examined the amendment in question and decided on a different approach: "The current definition of variola virus, as provided in the statute, could be interpreted to include other less harmful naturally occurring poxviruses such as vaccinia virus that are vital to beneficial research, thereby inadvertently restricting and potentially making criminal many types of beneficial research such as the development and production of smallpox vaccine. For these reasons, the NSABB recommends that [the amendment] be repealed, particularly because the misuse of variola virus is adequately covered by other criminal laws already in place."[22] Thus the primary advisory group convened by the U.S. government to guide federal biosecurity efforts has recommended repealing the first law limiting the use of DNA synthesis, which is an interesting start to an entirely new branch of technology policy.

The ability to synthesize a pathogen from scratch appears to many people to be a threat requiring legal prohibition. However, the combination of the flexibility of DNA-synthesis technology and the complexity of specifying the precise nature of the threat make it exceedingly difficult to draft clear and unambiguous legal prohibitions. New technologies are forcing policy makers and scientists alike to reevaluate what is possible in the lab, what should be considered a threat, and how to deal with that threat. Even more complicated issues will arise in the years to come.

Pending and Recommended Regulations

Restrictive regulations and the imposition of constraints on the flow of information are not likely to reduce the risks that advances in the life sciences will be utilized with malevolent intent in the future. In fact, they will make it more difficult for civil society to protect itself against such threats and ultimately are likely to weaken national and human security. Such regulation and constraints would also limit the tremendous potential for continuing advances in the life sciences and its related technologies to improve health, provide secure sources of food and energy, contribute to economic development in both resource-rich and resource-poor parts of the world, and enhance the overall quality of human life.[23]

So says a report from the National Academy of Sciences, written by the Committee on Advances in Technology and the Prevention of Their Application to Next Generation Biowarfare Threats, a collaboration of the National Institutes of Medicine and the National Research Council.[24]

At the end of 2006, the NSABB released its own initial report on synthetic threats. Among the recommendations made by the NSABB are those for (1) a specific definition of which sequences are covered by the Select Agents Registry, (2) a formal and consistent process for comparing synthesis orders to the registry by using software, and (3) the maintenance of records of orders for five years. The board noted that "effective compliance requires provider acceptance and may also require audits, fines and/or other legal actions."[25]

The Sloan Report

The NSABB notes that "the speed of advances in [gene-synthesis] technology will require governance options that are capable of keeping pace with rapidly evolving science." This is the immediate concern of a report released in the fall of 2007, *Synthetic Genomics: Options for Governance*, funded by the Alfred P. Sloan Foundation (hereafter referred to as "the Sloan Report" or "the report"): "The goal of the project was to identify and analyze policy, technical, and other measures to minimize safety and security concerns about synthetic genomics without adversely affecting its potential to realize the benefits it appears capable of producing." The authors of the report also note, "We found no 'magic bullets' for assuring that synthetic genomics is used only for constructive, positive applications. We did, however construct a series of policy interventions that could each incrementally reduce the risks from this emerging technology and, if implemented as a coordinated portfolio, could significantly reduce the risks."[26] The report identifies three major areas for policy intervention: (1) commercial firms that sell DNA (oligonucleotides, genes, or genomes) to users; (2) owners of laboratory benchtop DNA synthesizers, with which users can produce their own DNA; (3) the users (consumers) of synthetic DNA themselves and the institutions that support and oversee their work.[27]

Developing Policy Recommendations
beyond the Sloan Report

What follows is a brief examination of two facets of the Sloan Report that require further scrutiny: (1) the scope of the governance options is too narrow, and (2) a set of "legitimate users" is assumed in the absence of a definition of "legitimate."

In future discussions, the scope of options should be expanded to include today's baseline policies. Missing from options listed in the report is "do nothing," or even "maintain the status quo," which, because of the existence of the Select Agents Registry, is already more than nothing. The only options presented constitute new regulations of one kind or another. This shortcoming is compounded by the text immediately following the list of policy options, which reads, "This report presents no recommendations." Yet maintaining the status quo is not considered among the options for the future.

The text of the report is thereby an argument that intervention is required to improve security. By listing only policy options that go beyond today's baseline, the authors implicitly assert the need for *further* regulation. Through this assumption of the need for regulation and through the explicit identification of some presumed set of "legitimate users," for which the qualifications for legitimacy are not defined, the report in effect recommends restriction of access.

Potential Unintended Effects of Regulating DNA Synthesis

Implementing restrictions on biological technologies that might improve safety and security should always remain among our options. But implementing regulations without a careful examination of possible consequences is unwise.

Instituting security measures, and maintaining auditable records of both security and access, will incur costs for producers, users, the government, and society as a whole. Understanding the potential costs of restricting access first requires examining the proposed mechanism of regulation in slightly greater depth.

The conversation about regulating DNA synthesis often revolves around a set of "legitimate users." Unfortunately, it is not at all clear that regulation will limit access to synthesis technology by users who may be considered a threat. Restricting access to DNA synthesis may motivate some consumers—including those most deserving of scrutiny—to seek access to producers who are either not bound by restrictions or who are willing to ignore them.

In order to facilitate the control of access, the Sloan Report proposes the option of establishing a registry of DNA synthesizers, service providers, and certified users. Such requirements would allow DNA synthesis only in what amounts to secure facilities, where security is defined by monitored operation of DNA-synthesis technology either through licensed or permitted ownership of instruments or through licensing of "legitimate users," or both. Sequences submitted to these secure facilities might be kept on file for some number of years to facilitate any forensic efforts. Screening software would examine submitted sequences to identify potential threats in the form of genes and pathways that code for toxins or genomes that code for pathogens.

With respect to the costs of this registry, the Sloan Report notes that "if a review mechanism were too burdensome, small start-up firms might shift to in-house DNA synthesis instead."[28] Thus one of the immediate *social* costs of implementing a registry might be that some otherwise legitimate users opt out of participating due to the *monetary* costs of compliance, thereby limiting the utility of the registry. Those who choose to "drop off the grid" by synthesizing genes in-house could be monitored only if reagents and instruments were strictly controlled. As a result, one potential outcome of restricting access to synthesis might mirror the problem encountered by the U.S. Drug Enforcement Agency when it cracked down on domestic methamphetamine production (see Chapter 9): information on activities the agency wished to monitor and suppress became much harder to obtain. Similarly, regulatory actions that motivate users to pursue synthesis outside the registry may reduce knowledge of what is being synthesized, and by whom.

In this context one must also keep in mind the already intrinsically international nature of the DNA-synthesis market. Effective regulations on access to synthesis must therefore be international in scope and must track the flow of valuable design information through electronic networks, often across borders. This point brings us to the most important vulnerability in DNA-synthesis screening, one inherent in the inevitable and increasing reliance on information technology. Whether in the form of electronic signatures, databases of "legitimate users," screening software, or a design tool, this information can be viewed, copied, and even altered. Moreover, it is subject to a growing number of security threats that cover the range from simple human mistakes, to fraud, to interception of information during transmission, to complex software attacks on other complex software.

Under any international regulatory regime that required screening, individual firms would be faced with exposing their designs to multiple sets of eyes, which would threaten their economic security. As of 2006, industrial espionage was estimated to cost firms at least $200 billion annually.[29] Commercial risks due to exposing proprietary designs come in two forms.

First, it is not possible to guarantee any commercial firm that a design file, particularly if stored electronically for any length of time, would be secure from prying eyes or meddling hands. Second, interception of physical DNA sequences shipped from licensed providers would enable reading that DNA via sequencing instruments. Today, sequencing genetic circuits that have economic importance—all of which are much shorter than a full genome—would require no more than a few days of effort, and probably no more than a few hours, for well-equipped sequencing facilities (see Chapter 6).

In both cases, reverse-engineering a synthetic circuit of modest size from sequence data to infer function would not be difficult. This is true today because most genes (or variants thereof) that one might include in a new design are either already deposited in public databases or are closely related to genes in those databases. Currently, it is highly unlikely that there are sufficient secret genetic or metabolic data to serve as a barrier to reverse-engineering any synthetic circuit. It is only a minimal additional step (though potentially still expensive for the time being) to resynthesize and thus make use of intercepted sequences.

Over the coming years, the protagonists and antagonists (as defined by the reader) in the above paragraphs need not be using the same software, nor the same models and design tools, nor the same database of parts. Relying on software and models for security requires that these tools always be the best and that databases always be complete. Yet there will be constant competitive pressure to improve design and screening tools, in a context where innovation to produce better tools is possible anywhere in the world. This is not a game that can be won by drawing any sort of line in the electronic sand.

This "arms race" is an explicit indication that predictive design tools are required for future safety and security, while simultaneously being dual-use technology. These design tools can then also be used for espionage, industrial or otherwise, forcing designers to obfuscate their designs or seek noncompliant synthesis providers, further complicating the already challenging task of ensuring that only legitimate users have access to the technology. A model that can predict genetic mechanisms for the purpose of identifying potential threats, based on sequence information alone, can by definition be used to create those threats. Stated more directly, in the context of globally distributed biological technologies there will be a constant struggle to produce tools that better identify threats, a struggle in which those tools function well only because they enable threats of similar, if not greater, sophistication.

The Collision of Innovation and Regulation

My greatest concern is that we do not now have the technological capability to respond to rapidly emerging natural or artificial threats and that the development of such technology requires vastly more innovation than we have demonstrated in biology thus far. Rapid innovation in aviation and computation during the twentieth century was driven by (1) government investment to support the development of foundational engineering and production technologies and (2) a swarming mass of commercial innovators of various sizes who invent and produce the tools we actually use. If we are to encourage the widespread innovation necessary to produce real, commercially viable technologies—that is, something an individual or government can purchase and rely upon—then we must also arrange our security measures in a way that encourages, rather than discourages, commercial innovators to invest and risk both capital and reputation.

Given a choice—or if forced by regulatory action to make a choice—some designers of new DNA circuits will likely conduct business with synthesis providers who do not maintain an archive of design files. In the event regulations requiring such record keeping are implemented, in order to be effective, the relevant legislation must also specify that no domestic designers may contract with synthesis providers in countries that do not have similar provisions for record maintenance. While such legislation might pass and be coordinated internationally and might also pass muster with the World Trade Organization (though this would be tested in court, eventually), enforcement will become increasingly challenging.

Designers would face difficult choices. They would be asked to take a risk that record storage at a synthesis provider could be compromised, likely resulting in the loss of business and intellectual property (IP). Or they could take a risk that a plasmid smuggled through international shipping routes—perhaps simply via the mail—would be intercepted by authorities, which, though unlikely, would result in legal action, almost certainly including loss of business and IP.

In this situation the legal route poses much greater risk of compromised intellectual property than the illegal route. It is simply not clear that any enforcement methodology could effectively screen for smuggled DNA. As it stands today, completely innocuous DNA is often transported or sent through the mail between research groups in dried form on pieces of paper. The point is not that a great many designers or consumers of DNA synthesis services would be willing to risk jail to avoid losing IP but that anyone who wants to get DNA from offshore can do so via any number of routes, with nearly unlimited freedom to innovate and develop new ways to circumvent any prohibition.

Erecting effective barriers to physical transport of synthetic DNA would be extraordinarily expensive. This is just one more indication that regulating use and access to DNA synthesis may not be the most effective way of improving safety and security. Without an explicit evaluation of the commercial impacts of regulatory measures, and of the consequences of market-inspired cheating, implementing regulation is problematic at best.

A rush to regulate reflects the public discussion thus far. Academic and government policy recommendations tend to be made without a realistic consideration of how new technologies are developed and used. There is insufficient discussion of the market in which synthesis companies operate, how this market may change with increasing demand, and, most important, the behavior of consumers of synthetic DNA under various regulatory schemes. The simultaneous proliferation and centralization of methamphetamine production serves as a very clear counterexample to the assertion that safety and security can be found in the implementation of regulations on an already widespread technology.

The relevance of this observation to building a secure and robust bioeconomy is twofold. First, methamphetamine is a manufactured product, and attempts to constrain manufacturing have resulted in both greater production capacity and greater market opacity. As has always been the case in policing and intelligence work, information is the key to successful law enforcement or national defense, respectively. Therefore, looking forward to the bioeconomy, maximizing the free flow of information to authorities is crucial for security, perhaps even at the cost of allowing access to technology and skills by individuals who are the subject of concern by those same authorities.

Second, focusing regulation on the supplier rather than the buyer does nothing to alter demand but rather displaces production elsewhere and enhances substitution. From the perspective of security, this problem is exacerbated when the subject of the transaction has value beyond the buyer, as is the case in the marketplace for synthetic genes. In other words, there is a clear and growing demand for synthetic genes not for their own sake but for their value in producing materials and fuels that have even greater value, economic or otherwise. Thus the presence of any bottleneck, be it economic, technical, or regulatory, in turning electronic sequence specification into physical DNA will simply encourage alternate supply in an already international market. This dynamic is developing in the gene-synthesis market sooner than expected.[30]

As an example, the cost advantage held by a small number of synthesis companies funnels many orders to their foundries. But the extant structure of the proprietary gene-synthesis market is already causing some dissatisfaction among customers. In light of numerous informal conversations, it is already becoming apparent to me that parties interested in building synthetic genetic circuits or organisms are uncomfortable with exposing proprietary designs to scrutiny by any potential competitor or any third party with a potential conflict of interest. This would include all gene-synthesis companies that aim as part of their strategy to build design services atop their synthesis businesses. Furthermore, while multigene-length sequences are now often delivered within two or three weeks, this delay has already become the rate-limiting step for design cycles in synthetic biology companies. That is, the quest to field a marketable product is slowed by delivery times for outsourced DNA fabrication, which simultaneously requires exposure of proprietary design work and strategy to outside observers.

This effect is at least an implicit goal of some who consider centralized gene synthesis to be a security advantage. Low-cost, high-volume centralized foundries are said to allow for more-effective screening of orders for sequences of concern. The lowest-cost, least intrusive policy option identified in the *Synthetic Genomics: Options for Governance* report is for synthesis companies to retain records of orders for five years.[31] But the combination of IP issues and inefficiencies caused by lengthy delivery times will undoubtedly create a market for alternative synthesis technologies. As a result, I suspect there will soon exist, if it doesn't already, a market for desktop gene-synthesis instruments, even at a significant price premium.[32] These instruments would eliminate IP concerns and could provide significant cost savings (primarily in labor) as genes are produced in-house in days rather than weeks.

Where Do We Go from Here?

Every month, and every iGEM, brings news of synthetic systems of surprising complexity—systems that function more or less as intended. But the difference between "more" and "less" is the crux of many concerns about the future. The behavior of most synthetic biological systems is still difficult to predict, a state of affairs likely to persist for some time. Whether synthetic vaccines, genetically modified crops, or simple summer projects, systems composed of many poorly defined components, and their poorly defined interactions, will display unexpected behaviors.

Crucial to the future development of technologies used to manipulate biology is the explicit acknowledgement by policy makers, practitioners, producers, and consumers that biology is itself a technology. As such, biological technology requires a decision-making process based on the best available data to evaluate the wisdom of particular implementations. No bridge, dam, airplane, car, or computer is today built in developed economies without an evaluation of failure modes and consequent impacts. There are, of course, exceptions to this way of doing business. Risk factors and impacts may be ignored or overlooked, resulting in buildings or bridges that collapse, cars that tend to roll over or explode in collisions, genetically modified cotton plants whose boles fall to the ground for mysterious reasons, and gene therapies that cause rather than cure disease.

Design and construction standards for large and visible infrastructure are easy to police because it is hard to practice anything resembling civil engineering while incognito. Similarly, when cars or airplanes cost lives or when medical personnel cause injury, these red flags are relatively easy to spot, and those responsible tend to be held accountable according to the precepts of local legal systems.

In evaluating such consequences of bad decision-making and in drafting regulations or legislation to improve safety, we must distinguish between negligence and technological stumbles. There is a great difference between a plane crash resulting from failure of the air traffic control system, or mechanical failure of a faulty or poorly maintained part, and a crash resulting from phenomena that were not previously understood as important in the engineering process, such as metal fatigue or wind shear. Lives may be lost in all cases, but differentiating individual negligence from collective ignorance is a key role of existing criminal and civil justice systems, legislation, and administrative rules. It is by no means clear that *extra* regulation is needed in the case of biological technologies or that it would provide additional safety and security.

In general, there are two types of restrictions on actions in our society. The first type, coming into play before any action is taken, consists of limits on practicing certain skills, in the form either of legislation by the state or of professional certification. The second type comes in the form of remediation once action results in physical, economic, or social harm. It is unlikely that anyone interested in these issues or involved in relevant debates would argue against legal penalties for those responsible for a use of biological technologies that caused harm. But it is a very different style of regulation to restrict access to, or outlaw use of, biological technologies than to penalize harm to property or person resulting from negligent or premeditated use of those technologies. And it is by no means clear who should be subject to regulation or certification.

A case can be made that future engineers or artisans designing synthetic biological systems for health-care applications, or vaccines, or even some day for growing houses, should perhaps be required to "sign their drawings" as professionals. But this begs the question of what to do about hobbyists and do-it-your-selfers. While even those who remodel their own houses are subject to building codes, there are always places where codes do not exist, are not enforced, or are not enforceable. The lesson is that where access to tools and skills is ubiquitous, those who choose to do things their own way can always find a place, or a set of conditions, that allow them to express themselves or to experiment. The level of intrusive surveillance required to monitor everyone who wants to build glow-in-the-dark bacteria would in all likelihood be exceptionally expensive and is probably infeasible even within the United States.[33] As more BioBrick parts become available and as more people have general access to a combination of sequence specifications and synthesis, the task of enforcement resulting from restricted access or practice will become increasingly untenable. This returns the discussion to the dilemma of if and how to limit participation by entrepreneurs.

Regulation will *not* prevent the development of artificial biological threats and will *not* prevent accidents even among those authorized to pursue the development of supposedly benign applications. When harm does occur, we may never be satisfied that either criminal or civil penalties will constitute adequate repayment. Consequently, as a society we may yet choose to regulate access in addition to legislating penalties for harm. But there is no choice on the table that simultaneously provides safety through regulated access and provides technical innovation sufficient to address such issues as, for example, responding to infectious diseases, growing adequate and secure food supplies, and producing large volumes of biofuels. In contrast, regulation *will* impede innovation and therefore *will* impede our ability to respond to coming challenges. A choice to limit innovation explicitly opts for a reduction in capability and a reduction in preparedness.

The Futility of Prohibitions

"History shows that it is very hard to say no to technology."[34] Kevin Kelly, a founding editor of *Wired* magazine and longtime student of the role of technology in society, recently made a survey over the last one thousand years of large-scale technology prohibitions, defined as "an official injunction against a specific technology made at the level of a culture, religious group, or nation, rather than as an individual or small locality." While the results are thus far published only online and no source material or methodology is reported, Kelly's experience and stature as an observer of technology and its evolution provide what is—at minimum—a fascinating and important anecdote.

The results of the survey are remarkable for several reasons. First, Kelly could find only forty instances of prohibition that met his criteria, which means they are rare. Second, and perhaps most important for the present

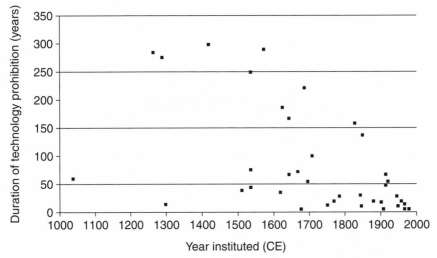

Figure 10.1 The duration of a technology prohibition plotted against the year in which it was inaugurated. Source: Kevin Kelly, "The futilitiy of prohibitions," 2006 [2 March 2009], available at http://www.kk.org/thetechnium/archives/2006 /02/the_futility_of.php.

context, prohibitions appear to be getting shorter. Figure 10.1 is a plot of the year a prohibition was instituted versus its longevity.

Kelly is a firm believer in the idea that all technologies, not just the biological technologies quantified in Chapter 6, are in some way improving exponentially. In Kelly's view, "as technology accelerates, so does the brevity of prohibition," a phenomenon exacerbated by ubiquitous access to information: "If we take a global view of technology, prohibition seems very ephemeral. While it may be banned in one place, it will thrive in another. In a global marketplace, nothing is eliminated. Where a technology is banned locally, it later revives to global levels."[35]

Even if the reader views this as a hypothesis rather than a firmly supported conclusion, it is a *strong* hypothesis in the sense that there is not only relevant data but also a theoretical argument that new technologies, even when unwanted by some people, often are put to economic advantage or otherwise exploited by others. When a technology becomes useful as an economic lever, objections tend to evaporate in the rush to apply that lever to raise one's own position in the world. We are at a point where a burgeoning industry is about to produce a great many more levers, many of which we require in order to improve our economic and physical security. It will not be a simple matter either to ban any given lever or to maintain that ban in the face of global innovation.

CHAPTER ELEVEN

Laying the Foundations
for a Bioeconomy

THE TITLE OF THIS CHAPTER IS, of course, behind the times: we already have a thriving bioeconomy. Without high-yield agriculture, the scope and accomplishments of human society would be severely limited, and without access to the fossil remains of prior life on Earth, now mined as petroleum, coal, and methane, we would be without considerable volumes of materials, fertilizer, and fuel and would be impoverished further still. Increases in agricultural productivity are just one example of improvements in biological technologies, which is particularly relevant here because the U.S. Department of Agriculture (USDA) claims that agriculture relies more on technology to achieve productivity growth than any does other sector of the economy.[1]

How Big Is the Present Bioeconomy,
and How Fast Is It Growing?

In public discourse the words "biotechnology" and "biotech" are often used in very limited and inconsistent ways. Those words may be used to describe only pharmaceutical products or, in another context, only the industry surrounding genetically modified plants, while in yet another context they refer to a combination of biofuels, plastics, chemicals, and plant extracts. The total economic value of biotechnology companies is therefore difficult to assess. It is a further challenge to disentangle the component of revenue from public and from private biotechnology firms.

Estimates of total revenues in the United States range from $200 billion to $250 billion annually, the specific dollar value depending on which set of products are included. The various surveys that provide this information differ not only in their classification of companies but also in their methodology, which in the case of data summarized by private consulting firms is not always available for scrutiny. Further complicating the situation is that data from private companies are self-reported, and there are often no publicly available documents that can be used for independent verification. One estimate, based on data from 2004, suggests that approximately 85 percent of all pharmaceutical "biotech" companies are private, accounting for a bit less than 50 percent of employment in the sector and 27 percent of revenues.[2]

These data are explored in more detail below, but a rough summary is as follows: As of 2007 biotech drugs accounted for about $79 billion in sales worldwide, with about 85 percent of that in the United States. Genetically modified crops accounted for about $128 billion, with 54 percent of that in the United States. Industrial applications (including fuels, chemicals, materials, reagents, and services) contributed another $70 billion to $100 billion in the United States, depending on who was counting and how. Annual growth rates over the last decade appeared to be 15–20 percent for medical and industrial applications and 10 percent for agricultural applications. After sifting through many different sets of numbers, as of late 2008, I estimate that "biotech" revenues within the United States were about $220 billion—economic activity that was the equivalent of nearly 2 percent of GDP—and were growing at a rate of 15–20 percent annually (see table 11.1). The U.S. GDP was about $14 trillion in 2007 and grew at an estimated 2.2 percent.[3]

Health-Care Biotech

Pharmaceuticals can be grouped very roughly into two categories. "Small molecules" are chemical entities and are generally produced via chemical synthesis for human use. "Biologics" are proteins or nucleic acids generally produced within cell culture systems or purified from plants or animals, and the production organism is frequently genetically modified to improve yield. Small molecules come, in other words, from putting chemicals in a pot and stirring, whereas biologics come from organisms.

The pharmaceutical industry overall amounts to about $250 billion in sales annually in the United States and $600 billion worldwide, with a re-

TABLE 11.1   Revenues from genetically modified systems in 2007

Sector	Worldwide revenues ($ billions)	U.S. revenues ($ billions)	% of U.S. GDP (total of ~$14 trillion)	Revenue growth rate in U.S. (%)
Biotech drugs ("biologics")	79	67	.48	15–20
Agbiotech/ GMOs	128 (est)	69	.49	10
Industrial	110	~85	.61	15–20

cent growth rate of 6–8 percent in the United States and of 81 percent in emerging markets.[4] Estimated sales of biologics in the United States range between $50 billion and $67 billion in sales in 2007, with a 20 percent annual growth rate over the preceding five years.[5]

The rest of the pharmaceutical market—the small molecules—depends heavily on biological technologies during drug development and clinical trials. It could be argued that because developing new small-molecule drugs requires the use of biotech tools, the entire drug industry should be included in estimates of the contribution of biotech to the overall economy. Most drug candidates are now tested first in model systems, beginning with single cells and then slowly moving up through model organisms such as mice, rabbits, ferrets, dogs, and primates. The effects of drugs in those systems are measured via many different molecular assays: some monitor the physical state of DNA, some monitor gene expression, and others monitor the amounts of proteins and their interactions or levels of various metabolites.

While the precise fraction of drug sales dependent upon biotech tools is probably a closely watched statistic within pharmaceutical companies, it is difficult to assess from the outside. If all small molecules on the market today are the product of a development and testing process that relies on biological technologies (almost certainly an overestimate), this would add $250 billion to the portion of U.S. GDP contributed by industrial and agricultural biotech, for a total of about $470 billion or almost 4 percent of GDP. But even if only half of small-molecule sales can be attributed to relying on biological technologies, it still starts to look like serious money.

Estimating the future health-care revenues from biotech is complicated, to say the least. Biological technologies could in the future have significant impacts both on the capabilities of health-care practitioners and on the revenues of the industry. The ability to use rational engineering to produce biologics and replacement tissues could, depending on cost and accessibility, transform the human condition. But there are many nontrivial issues obstructing a simple analysis.

Throughout this book, I have generally left unexplored the many applications of biological engineering to human health care. The straightforward reason for this omission is simply that the area is extremely complex from every perspective, including the actual molecular mechanisms of drug action, the metrics for effectiveness used in clinical trials, the manufacturing processes, cost-benefit analyses of adopting a new drug, drug approval and regulation, and marketing to physicians and patients.

Even in the high-growth market for biologics, it is easy to stumble. Pfizer recently withdrew its inhalable form of insulin from the market because of extremely weak sales, despite what seemed at the outset a significant advantage over injectable forms of the drug. Unfortunately for Pfizer, patients didn't like the inhaler, the drug cost more than alternatives, and insurance companies were not enthusiastic about paying nearly twice as much per dose, all of which contributed to sales of only $17 million in the face of projections of a billion dollars. Thus Pfizer decided to take a $2.8 billion charge by canceling production and distribution.[6] This is an indication that the products of high-end biotechnology, even when aimed at large markets such as diabetics, are already becoming commodities in a very competitive market, subject as all commodities are to price differences and choices made by consumers.

As biological production technologies mature, competition will only increase as new methods propagate and new producers enter the market. Off-patent biologics could well represent an even bigger challenge for pharmaceutical companies than generic small-molecule drugs do; generic and off-patent drugs reduced branded drug sales by $18 billion in 2006 alone.[7]

The pharmaceutical industry is suffering in part because companies are trying to address an intrinsically hard problem. Attempting to repair the human body without engineering manuals is no easy task. Approvals of new small-molecule drugs have fallen by about 30 percent over the last decade, despite a doubling in R&D spending and increased identification of new candidate drugs.[8]

One strategy in the face of declining new-drug approval is to focus on the segment of the population in which drugs have a higher likelihood of being effective, an emerging branch of health-care called "personalized medicine." This tailoring of treatment to the individual relies on the field of "pharmacogenomics," which aims to tailor therapies according to an individual's genetic makeup.[9] Beyond determining treatments based on a patient's genome lies "theragnostics," which is the fusion of therapeutics and diagnostics and focuses not solely on genetic variation, "but rather on the integration of information from a diverse set of biomarkers (e.g., genomic, proteomic, metabolomic)."[10]

The desire and apparent need to tailor treatment and drug regimes to individuals help explain a recent trend in acquisitions. Pharmaceutical companies are buying firms that provide diagnostic tools and services. Roche spent almost $4 billion on such acquisitions in 2007 alone.[11] This phenomenon is also reflected in drug sales, with drugs aimed at specific mechanisms (such as interfering with HIV proteases or increasing production of red blood cells rather than "reducing inflammation") increasing their share of revenue growth from one- to two-thirds between 2000 and 2006. Only 25 percent of new drugs are targeted to chronic, ongoing diseases like diabetes or high blood pressure, "suggesting the pipeline is shifting toward targeted therapies."[12]

Agricultural Biotech

Based on the U.S. Census Bureau Statistical Abstract, the value of agricultural production in the United States was about $300 billion in 2006.[13] Another estimate placed this value much higher: at least $800 billion per year as of 2001.[14] The recent run-up in commodities prices would push this higher still, though such data is not yet available. Genetically modified (GM) crops are still a relatively small fraction of total revenues. As of 2007, 114 million hectares of GM crops had been planted worldwide, on about 9 percent of the worldwide total for cultivated land, with a worldwide value of more than $100 billion. GM acreage has been growing globally at just over 10 percent each year for the last decade, with 54 percent of GM crops planted in the United States; in 2007, among staples worldwide, GM crops accounted for 61 percent of corn, 83 percent of cotton, and 89 percent of soy.[15] In the United States, in 2007, GM crops accounted for 73 percent of corn, 87 percent of cotton, and 91 percent of soy, for a total value to farmers from those three crops alone of about $69 billion.[16] China increased its planted acreage of GM crops by 20 percent just between 2006 and 2007.[17] Worldwide revenues from GM seeds were about $8 billion in 2008,[18] and could rise as high as $50 billion by 2025.[19] Brazil, Argentina, and Canada are similarly planting increasing amounts, and fractions, of GM crops.[20]

While most existing GM crops are modified with a single gene altering a single trait, the next generation will contain multiple genes that through their interactions confer more-complex traits. Drought tolerance in cereal crops is one such desired trait. Water shortages have caused Australia's farm output to fall by as much as 40 percent in recent years. Losses in the United States in any given drought year can be in the range of $5 billion to $6 billion.[21] The cost of developing drought-resistant crops is expected to be so large that even the biggest agricultural biotech companies are partnering up to share risk and financial burdens: Monsanto has paired up with BASF, and Dupont-Pioneer has signed deals with Dow and Syngenta. The demand for grain is also clearly very large, with increasing imports into Asia and a significant fraction of worldwide agricultural yield dedicated to growing energy crops. Multinational agbiotech companies are betting that they can supply this demand with GM grains. But the demand for GM crops is not uniform around the globe.

The general embrace by U.S. farmers of GM crops and the contemporaneous rejection of those crops by European consumers are producing yet another complexity within markets. While the European region is presently a net food exporter, much of the feed for livestock and poultry in the region comes from the Americas. Yet the strict safety testing and labeling requirements for food or feed containing GM plants amounts to a European zero-tolerance policy for importation of GM products. While GM sugar beets and some varieties of GM corn may be officially approved for sale in Europe, consumers appear to avoid products with the GM label. This policy has fascinating secondary consequences, namely, that it is on track to force dramatic reductions in European livestock production due to increasing fractions of GM feed grains, as Peter Mitchell notes: "If a solution isn't found, European farmers will be forced into wholesale slaughter of their livestock rather than have the animals starve. Europe will then have to import huge quantities of animal products from elsewhere—ironically, most of it from animals raised on the very same GM feeds that Europe has not approved."[22] Changes in attitude that produce a marketing environment friendlier to GM products may alleviate this problem. Yet consumer resistance to GM products in Europe is both deep and broad. Even in the face of economic hardship, brought on by reduced food exports and increased domestic prices, consumers and interest groups may take many years to change their minds, if ever.

The complex interaction of consumers, interest groups, and regulators poses significant risk to companies developing GM crops. Moreover, in addition to dealing with outright opposition to their products, such companies face additional financial risk associated with court decisions regarding the approval process, after products are already on the market. In the spring of 2007, Judge Charles Breyer of the U.S. District Court revoked the 2005 USDA approval of an herbicide-resistant alfalfa on the grounds that the agency had violated federal laws by failing to sufficiently examine the environmental and economic impact of the GM strain. Alfalfa comprises an $8 billion market for U.S. farmers, with 17 percent of that exported to countries not entirely comfortable with GM crops. Reexamination of prior approvals is a broad concern for agbiotech companies, as one commentator notes: "The [Roundup Ready alfalfa] ruling is one in a recent series from federal courts and USDA that speak not only to the volume but also the precarious legal state of current GM crop-related activities."[23]

Some context for the potential risks of releasing GMOs into the wild can be found in estimates of the cost of invasive species. As of 2005, the total cost of invasive species in the United States was conservatively estimated to be $120 billion, including control costs and direct damage to crops, property, human and animal health, and water supplies. It is interesting to note that roughly 98 percent of the total annual value provided to the U.S. food system is from certain other nonnative species, including corn, wheat, rice, cattle, and poultry. The authors of the estimate note, "If we had been able to assign monetary values to species extinctions and losses in biodiversity, ecosystem services, and aesthetics, the costs of destructive alien invasive species would undoubtedly be several times higher than $120 billion/year. Yet, even this understated economic loss indicates that alien invasive species are exacting a significant toll."[24] Many of the species included in the damage estimate were originally introduced into the United States in the hopes that they, too, would contribute to the economy, demonstrating that in yet another area humans have a long way to go in understanding the effects of manipulating the biological systems in which we live.

Industrial Biotech

The size of industrial biotechnology is the most problematic to assess, as portions of it might be attributed either to pharmaceutical biotech or to the chemicals industry. Thanks to the biofuels boom of the last few years, industrial biotech is also growing rapidly. Moreover, at least to the extent that most production in this category is not intended for use as drugs and, as often as not, uses non-GM crops as feedstocks, the regulatory burden is much lower. From laundry enzymes to nutritional supplements to bioplastics to enzymes and organisms producing biofuels, the products of industrial biotech are pervasive in the U.S. economy.

Estimates from McKinsey and Company are that industrial biotech, "counting products made from biobased feedstocks or through fermentation or enzymatic conversion," amounted to U.S. sales of $50 billion in 2003 and $77 billion in 2005 and about $85 billion in 2007.[25] Much of that dramatic jump is attributed to an increase in biofuels production. The revenues in 2005 represent 7 percent of sales within the broader chemicals industry, and McKinsey forecast in 2006 that sales will rise to 10 percent for that industry by 2010, amounting to $125 billion. This figure is an adjustment down from the 2004 estimate of $160 billion in sales, representing 20 percent of chemical industry revenue.[26]

Summary of Bioeconomy by Sector

This backdrop sets the context for today's increased reliance upon genetic modification of biological systems for the production of food, drugs, materials, and fuel. Amid today's increasing worldwide fraction of GM crops, biofuels, and materials, we are clearly beginning to use biology in new ways. The relative contribution of the different sectors to the total is worth considering. Until recently drugs dominated "biotech" revenues in the United States, but now this contribution accounts for less than half the total. This suggests that as biological technologies mature, becoming more useful and prevalent across different sectors of the economy, industrial and agricultural applications will amount to a much larger share of total revenues. Moreover, while most health-care and agricultural applications are subject to regulation of one kind or another, most industrial applications are not. That is, the sector dependent upon biotech that constitutes the fastest-growing, and now largest, contribution to the economy is generally not subject to regulation.

Recall that the industry as a whole now employs about 250,000 people within the United States.[27] This skilled workforce represents only one-sixth of 1 percent of total U.S. workers while accounting for 2 percent of GDP and therefore contributes disproportionately to the U.S economy.

It is unclear what proportion of any country's GDP that biological technologies could ultimately provide. Biology is just one means to produce goods and must compete against existing and future technologies that might be better suited for certain purposes. However, there is considerable potential for economic growth via the development and deployment of biological technologies throughout the economy.

The earliest and most obvious examples of the coming economic impact of biological technologies are in the areas of biofuels and materials. As a commodity, biofuels provide an example of one end of the spectrum of biological products. Whereas biologics (biologically derived drugs) are produced in small volume with very high value, perhaps many millions of dollars per kilogram, biofuels will be produced in extraordinarily large volumes at both low cost and low margin. Moreover, biofuels production will require copious amounts of feedstocks. In 2006 and 2007, governments around the word set mandates for biofuel usage that not only require significantly increasing the amount of feedstock dedicated to energy production but that cannot be met without the introduction of new technologies.

Toward Building Biofuels

Technologies to produce biofuels are likely to come into the market in three phases over the next decade. The specifics of this story depend both on the continuation of present technological trends and on progress in areas of research and development that cannot be easily predicted. Nonetheless, it is possible to say something relatively concrete about the present state of the market and about the potential for new sources of fuels.

First Phase: The Collision between Food and Fuels

The present market for biofuels is characterized by a link between crops used as feedstocks and the use of those crops as food or animal feed. Current corn prices are an indication of this link: many analysts attribute recent price increases to a greater use of corn to make ethanol as liquid fuel.

The general phenomenon behind the coupling of grain prices and biofuels prices is not the use of grain to make biofuel per se but rather the climbing overall demand for grains. As a result, the grain surplus characterizing markets for the last several decades is now over and is not projected to return anytime soon (see figure 11.1).[28] Moreover, dramatic reductions in worldwide grain stockpiles, which might otherwise serve as a supply buffer, suggest that "overall price variability and market volatility in the agricultural sector are likely to increase."[29]

Despite common claims that rising prices are the result of overeager processing of food into fuel, the real situation is more complex. While the use of corn to make ethanol appears to be contributing to rising food prices, this cause-and-effect relationship has been inverted for palm oil.

In 2005 and early 2006, oil palms were seen as an excellent source of vegetable oil for conversion into biodiesel. European, American, and Asian investors poured money into facilities capable of refining that oil in the hopes of capitalizing on growing government mandates for blending biodiesel into the fuel stream. At the time these investments were made, there

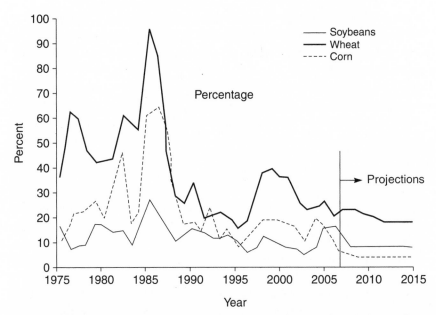

Figure 11.1 The ratio of stocks on hand of corn, wheat, and soybeans to their yearly consumption. Grain stocks are projected to remain at historic lows for many years to come.

appeared to be a substantial financial advantage in selling biodiesel, given the rising price of petroleum-based fuels.

Yet as of mid-2008, raw palm oil traded at a price approximately 30 percent higher than finished wholesale diesel. Prices climbed sharply during 2006 and 2007, due primarily to substantial increases in demand for palm oil in food markets in Asia.[30] This makes difficult the proposition of purchasing palm oil on the open market for the purpose of producing biodiesel. Significant refining capacity planned in 2005 and 2006 was in fact never completed in the face of rising palm oil prices.[31] Facilities around the world sat idle, or operated at a loss to fulfill contracts signed when palm oil refining could produce fuel at a healthy margin. Whatever the future of palm oil prices, the tension between food and fuel use of this resource will remain part of the market.

The collision between food and fuel will be maintained as long as there is a competition for feedstocks. Some relief will come from increasing yields from standard food grains. The average corn yield in the United States more than tripled between 1950 and 2000.[32] With an overall average of about 160 bushels per acre on functioning farms, test plots in the United States are now approaching 300 bushels per acre. Similar improvement should gradually emerge in other countries. For example, Chinese average yields in 2006 were approximately 85 bushels per acre, and there is every reason to expect new strains of corn planted in China to continue climbing the yield curve. However, while corn supplies may increase, demand may continue to grow faster. In 2008 the Chinese corn supply was expected to grow by 1.4 percent, while the demand was expected to grow almost 3 percent.[33] A deficit in domestic supply led China in December of 2006 to impose a moratorium on converting additional amounts of corn to ethanol.[34] As a policy matter, the government has apparently decided that the use of corn as food and feed is more important than its use as a feedstock for fuel.

Government targets for ethanol usage around the world are so high that there could be serious supply issues in just a few years. Warren Staley, the CEO of Cargill, put it this way: "Unless we have huge increases in productivity, we will have a huge problem with food production. The world will have to make choices."[35]

The global lack of surplus food and feed grain means the biofuels industry inevitably must look for alternative feedstocks in order to meet increasing demand. In this context considerations of the utility of a feedstock must therefore be broadened, somewhat unconventionally, to include land and water that can be used to grow "dual-use" crops such as corn, soy, and wheat. There exists a large amount of "infertile" land in the United States and around the world, i.e., land that farmers choose not to cultivate because it cannot be used to produce food at market prices but that might be used to grow biofuels feedstocks.

Consumers, too, make choices to use land and water in particular ways when deciding to eat certain foods. A kilogram of pork requires an input of three kilograms of grain; each kilogram of beef requires eight kilograms of grain.[36] The question human societies must debate, then, is *which* food to grow, and where.

Populations are distributed quite differently than either arable land or food production. According to both political and economic orthodoxy, those differences in supply and demand are supposed to be addressed by global, open trade networks. When the market fails, governments and international organizations typically step in and supply "aid." However, the use of fruits of the land as industrial feedstocks will further complicate choices about resource allocation and cannot help but increase prices for food, at least in the short term. Annexing more land into food and feed production adds further complexity, and exploiting increasing amounts of presently undeveloped natural resources will come at a cost.

Policy choices have a clear effect on investment and subsequent market dynamics. Mandates for ethanol usage in the United States have led to more than enough production capacity. Between the beginning of 2005 and late 2007, the number of plants in operation in the United States increased by more than 50 percent to around 130, while the number under construction more than quadrupled to about 80.[37] The resulting volume of ethanol on the market has depressed prices, leading to a recent reduction in investment and plans for new plants. The present ethanol oversupply will in turn ease upward pressures on corn prices, for the time being, but competition for all grains will resume eventually with the inevitable increase in demand for both food and fuel.

Second Phase: The Emergence of Dedicated Energy Crops, New Fuels, New Processing Technology, and the Debate over Carbon Emissions of Biofuels Production

Rising incomes around the world will continue to put pressure on commodities prices, as more grain is used for food and feed. Water and arable land are the primary resources required to provide that food and feed. Thus, in order to reduce competition between food and biofuels for resources, which would result in a decoupling of food prices from biofuel prices, biofuels producers must find crops for use as feedstocks that need not compete directly for land and water with edible food crops.

New energy crops are already being planted in many countries around the world. China and India are experimenting with large plantations of jatropha, an oilseed-bearing bush that grows in arid conditions and on poor soil. Jatropha seeds are composed of up to 40 percent oil and must be harvested by hand, then pressed to obtain the oil, and then the oil must be processed into biodiesel. The processing and infrastructure is no different than that required for palm oil. Because jatropha oil is not edible and grows in poor soils, it represents a feedstock that may begin to alleviate concerns about competition with food. Recognizing this potential, BP has invested $90 million in a venture with D-1 Oils to produce biodiesel from jatropha.[38] China is planning to plant up to 13 million hectares (mha) of jatropha by 2020.[39] India has earmarked at least 11 mha for growing jatropha, amid public discussion of up to about 32 mha, with the resulting diesel intended both for domestic use and for export to Europe.[40] However, because there are presently only a few plantations in commercial operation, and there is thus little experience with large-scale cultivation and harvesting, it is not yet clear what role jatropha will play in the global biofuels marketplace.

Similarly, a great many hopes are being put on cellulose as a feedstock for biofuels. In short, cellulose and hemicellulose, when bound with lignin, form the primary structural stuff that holds up green plants.[41] Cellulose is an extremely tough polymer consisting of sugar molecules very tightly bound together, while hemicellulose has a similar composition but is more easily broken down. Lignin is a non-sugar-based polymer. Whereas a crop like corn or wheat is presently grown for seeds—the kernels, or grains—to serve as food, feed, and industrial feedstock, these seeds account for only a fraction of the plant. Depending on the species, 50 to 85 percent of the mass of plants is composed of cellulose and hemicellulose, and 10 to 25 percent of plant mass is lignin. Thus, if the sugars comprising cellulose could serve as feedstocks for fermentation, the stalks of corn and wheat, as well as waste materials and the bulk of other plants and trees grown specifically for the purpose, might serve as a significant source of liquid fuels.

There are direct industrial routes to converting cellulose and other biomass directly to liquid fuel. At high temperatures, in the absence of oxygen, organic material decomposes into simpler components making up "synthesis gas." A similar result can be obtained via acid or steam treatments. With the right catalyst and careful process control, synthesis gas can be recombined to produce ethanol and other compounds. This technology is relatively old, dating to the end of the nineteenth century, but because of the high energy cost of thermal decomposition and the complicated chemistry of both decomposition and catalytic synthesis, the approach has until recently proved too expensive for use in making liquid fuels.

Beyond ethanol, there is presently growing interest in butanol as a transportation fuel. Because butanol is widely used as a solvent, the industrial production process has been under constant pressure to reduce costs for almost seventy-five years. Yet the existing process results in a product approximately 20 percent more expensive than gasoline in the United States. Consequently, there are considerable efforts being put into developing biological routes for butanol production. Butanol has important physical advantages over ethanol, including a higher energy content per unit volume and a much lower affinity for water.

Butanol is an "old" transportation fuel; U.S. farmers produced it during the early mid-twentieth century for use in tractors. Just as yeast ferments sugar in ethanol, so the bacterium *Clostridium acetobutylicum* ferments sugar into a mixture of acetone, butanol, and ethanol. The natural yield of the organism is relatively low in butanol content. However, the relative physical advantages of butanol over ethanol have not escaped the notice of biofuel producers today.

In 2007, BP and DuPont plan to begin introducing butanol as a fuel-

blending agent in the United Kingdom by 2009.[42] The pair of industrial giants, along with British Sugar, will initially produce about 35 million liters
annually from sugar beets by using existing technologies. The firms also
plan to invest $400 million in building a 420-million-liter-capacity plant in
the United Kingdom, scheduled to be commissioned in 2010. This plant
will initially produce ethanol from wheat but will be converted to butanol
production with the introduction of a "generation 2 biocatalyst"—a modified microbe—that produces butanol at a higher yield. DuPont expects to
introduce this microbe by 2010, resulting in production of butanol at the
equivalent of $30–$40 per barrel, at which time it also expects to begin
marketing butanol in the United States.[43]

Biofuel production will also benefit from the use of enzymes that assist
in the breakdown of cellulose and starch. These enzymes might be derived
from the bacteria that live in termite guts or from the bacteria that live
within ruminants. Enzymes can eliminate the "cooking" step for processing of high-starch feedstock, by directly digesting granular or uncooked
starch to create fermentable sugars, or to aid liquefaction of corn solids, or
to help clean up vegetable-oil mixtures for processing into biodiesel. Verenium is collaborating with the agricultural biotech firm Syngenta to genetically modify corn to express another enzyme, Amylase-T, which would
digest starch into sugar within the growing corn plant, potentially providing a significant reduction in subsequent processing energy and cost. As of
mid-2009, the collaborators are awaiting approval from the USDA for
field trials. Genencor, Verenium, and Novozymes, among a raft of other
companies, are developing cellulase enzymes that help break down cellulose.

To aid in commercializing cellulosic ethanol, the U.S. Department of Energy (DOE) in early 2007 announced government funding for six pilot bio-
refineries.[44] The DOE funding of $385 million is reported as leveraging
considerable private funds for a total investment of $1.2 billion. Using var-
ious combinations of acid treatment, steam, enzymes, and fermentation,
the pilot plants are expected to produce 130 million gallons of ethanol annually from wood, yard waste, corn stover (the leftover stalks), wheat straw,
and other feedstocks.

These technologies may reduce the cost of processing corn into ethanol,
but this strategy attacks only a small part of the problem today: most (75–90
percent) of the cost of biofuels is in the feedstock. Fuels are commodities,
and the probability of ever cutting total cellulosic ethanol cost in half is
slim to none. As long as corn, or any crop that can be grown on land that
would otherwise be used for corn, serves as a feedstock, the profit margin
on the resulting fuel will be squeezed from below by food prices and from
above by both petroleum prices and competing ethanol suppliers.

There are two final problems with a policy of reliance upon cellulosic ethanol as fuel. First, there is no existing means in the United States to gather large amounts of corn stover, wheat stalks, or other biomass from fields and transport it to processing facilities. Except in limited circumstances, where waste from lumber mills, or municipal yard waste, is already concentrated at a particular location, considerable investment will be required to build and maintain the appropriate shipping infrastructure.

Second, while there are estimates for the performance of various cellulose-processing technologies, none of them have yet been operated at scale. This means there is uncertainty not only in the total output from any given plant but also in (1) the total number of plants necessary to provide a given volume of fuel, (2) the total volume of feedstock and total amount of energy necessary to run the plants, and (3) the total cost of the infrastructure to harvest and transport that feedstock. The latter point is particularly problematic, because while transportation costs depend linearly on the distance that feedstocks and finished fuels must travel, the total feedstock volume depends on the *area* that is harvested and therefore upon the *square* of the distance feedstocks must travel. Thus small changes in transportation costs can have large impacts on the volume of feedstock that can be economically processed.

A major motivation for increased biofuels production is reducing greenhouse gas emissions presently caused by burning fossil fuels. Producing large volumes of biofuel will require large-scale sources of biomass. The choice of crop is, of course, connected to where it will be grown, how it will be processed, and how far the resulting fuel must travel to reach consumers.

The issue becomes yet more complex when the total carbon emissions of farming are figured into these calculations. The overall carbon content of soil is many times that of the atmosphere. In addition to emissions from running farm machinery with hydrocarbon fuels, a significant fraction of the carbon emissions from standard agricultural practices come from planting and harvesting. Simply turning over the soil annually releases large amounts of carbon into the atmosphere because it breaks up root structures and destroys a subsoil ecosystem composed of microbes and fungi. One effort to integrate soil carbon measurements with an agricultural model of the north central United States found that the organic carbon content of the soil had decreased by around 20 percent since the introduction of agriculture into the area in about 1850.[45]

Undisturbed soil accumulates biomass over time. Switching to practices that do not involve turning the soil, so-called no-till agriculture, or substantially reducing tilling (conservation tilling), would result in a significant lowering in the total carbon emitted by farming and therefore by biofuel production.[46]

Whether grown via no-till or conventional techniques, high-yield GM crops are likely to become significant biofuel feedstocks, which will raise an altogether different set of problems. Leakage of genes from GM crops into their unmodified cousins is potentially a threat if herbicide-resistance genes are transferred into weeds. Gene flow into close relatives has been observed in tests plot of Kentucky bluegrass and creeping bentgrass, which provided "the first evidence for escape of transgenes into wild plant populations within the USA."[47] A similar result has now been demonstrated for a stable and persistent transfer of an herbicide-resistance gene from the widely cultivated *Brassica napus,* commonly known as rape or rapeseed, to its wild relative *Brassica rapa.*[48] Within the confines of a laboratory, herbicide-resistance genes can be transferred with relative ease via pollen exchange between common weed species.[49] These demonstrations may give pause to both policy makers and commercial interests. Any gene transfer in open cultivation that results in unintentional propagation of a new herbicide-resistant weed strain has the potential to cause substantial economic and physical damage.

The resulting potential threat to agricultural systems raises significant questions about the wisdom of relying on genetically modified crops for feedstock production. There is also a very real possibility of greater governmental regulation to limit planting of GM crops, or at least more-stringent application of existing regulation, if the ruling on GM alfalfa is an indication of future events.

Although the U.S. market has been relatively welcoming of GM crops, it is not clear that this attitude will continue. Energy crops modified to increase growth rates or to ease processing of cellulose and other components may face resistance based on the risk of such traits being transferred to wild plants and trees.

One potentially less controversial source of feedstocks is non-GM grass. David Tilman, an ecologist at the University of Minnesota, and his colleagues argue that restored native perennial grasslands could be used as a source of biomass for producing liquid fuels. The strategy essentially involves letting the North American prairie return to its native state and mowing the resulting grassland once a year to harvest biomass. More significantly, even using existing industrial technologies, the overall life cycle of these fuels might be carbon negative: that is, they might store more carbon than would be emitted during harvesting, processing, and use as fuel.

During a decadelong experiment, Tilman and his collaborators planted combinations of eighteen different native plant species in "highly degraded and infertile soil." The study, carried out using hand harvesting and weeding of small test plots, appears to demonstrate that biomass yields in diverse ecosystems exceed those from monoculture or low-diversity ecosystems.[50] Both total harvested biomass and sub-soil carbon increased as the test plots matured.

Within North America, several individual species of grasses are frequently suggested as candidates for large-scale *monoculture* cultivation as biofuel feedstocks. Switchgrass is one such example. Tilman et al., however, assert that while switchgrass may be highly productive in fertile soils, namely, those that can also be used to grow food, a monoculture of switchgrass growing in infertile soils is only one-third as productive as grasslands.[51]

This is in sharp contrast to a five-year study carried out by Marty Schmer and colleagues at the University of Nebraska. They conducted "field-scale farm trials" of switchgrass, planted on "marginal cropland" on ten farms and harvested using conventional hay-bailing equipment. The researchers assert their methods would produce almost twice as much estimated ethanol per hectare as prairies and nearly six times as much as other switchgrass trials.[52] Notably, Schmer and his colleagues found that with farm-scale harvesting, *actual* agricultural inputs—including diesel fuel, fertilizer, and herbicides—were as much as 65 percent lower than prior *estimates* based on small-scale tests.

Miscanthus x giganteus is another grass frequently touted as a cellulosic feedstock. "Giant miscanthus," as it is also known, is a sterile hybrid that is propagated by planting rhizomes—root structures—clipped from existing plants. In the first direct comparisons of *Miscanthus x giganteus* and switchgrass grown on "prime agricultural land" in the state of Illinois, starting in 2002, miscanthus produced twice the biomass per unit area as switchgrass.[53] Miscanthus is not native to North America, which might raise concerns about its potential spread as a weed, but advocates of miscanthus as a biofuel feedstock argue that it has not become invasive during more than two decades of cultivation at a test station in Denmark.[54]

In light of all these options, large-scale cultivation of grasses appears to offer real potential, subject to the rather significant caveat that we can presently do no better than estimate the total amount of any biofuel that can be made from any crop. Altogether, Charles Wyman, at the University of California, Riverside, estimates that "with 5–15 dry tons per acre per year possible for poplar, switchgrass, and miscanthus," the United States could entirely replace its gasoline consumption with ethanol, using 200 million acres of land, about 20 percent of the combined crop, pasture, and range lands.[55]

Nonetheless, the existence of an ongoing debate about which feedstocks might be best for growing biomass, and at what cost, serves to emphasize the early stage of knowledge and planning. The consequences of large-scale bioenergy production on carbon emissions, in particular, are in the very earliest stages of research. The first comprehensive studies, published together in *Science* in early 2008, are not encouraging. The authors of one paper concluded that converting native forests and grasslands in Brazil, southeast Asia, and the United States to grow biofuels feedstocks would release "17 to 420 times more CO_2 than the annual greenhouse gas reductions that these biofuels would provide by displacing fossil fuels."[56] Similarly, "if American corn fields of average yield were converted to switchgrass for ethanol, replacing that corn would still trigger emissions from land-use change that would take 52 years to pay back and increase emissions over 30 years by 50 percent."[57]

These results generated discussion that grew over many months in the pages of *Science* and other journals, with various academics, government scientists, and venture capitalists joining in. The "official" response of several U.S. government laboratories was to question the assumptions, methods, and motives of the original authors: "[The two papers,] widely reported in the press, raise important issues but often read like conclusions looking for an underlying rationale. These two studies fundamentally misunderstand the local forces behind land use change issues and make no provision for mitigating impacts such as the slowdown in urbanization that a vibrant agricultural economy would bring."[58]

The debate will obviously continue into the foreseeable future. It is too early to pass definitive judgment on any particular approach or any particular study. One thing is certain: if we are to make rational decisions, we require a much better understanding of the energy, carbon, and nutrient balance of various biomass crops when they are used to produce energy or materials. This consideration returns to the fore the issue of choice and public policy in determining the future uses of biology as technology. In this case, maximizing the yield of biomass and the volume of subsequent biofuels production is not necessarily the same thing as maximizing physical and economic security.

Overshadowing all other factors in using different kinds of land for different kinds of crops is the dramatic influence climate change may have on what can be grown, and where. Moreover, the total amount of land available for high-yield production of food may not be stable over time; according to the *Economist,* "by some measures, global warming could cut world farm output by as much as one-sixth by 2020."[59]

Third Phase: Production of Fuels Using
Engineered Microbes

With large-scale feedstock production still subject to market and regulatory maneuvering, the infrastructure to convert that feedstock into fuel is likely to change dramatically. Over the next five to ten years, fundamentally new biological technologies to produce biofuels will begin entering the market.

The fermentation of sugar to produce ethanol and butanol will be short-term solutions. The strategy of improving the biofuels-production pathways in existing organisms will rapidly be supplanted by new organisms, modified via metabolic engineering and synthetic biology, that directly convert feedstocks into transportation fuels similar to gasoline. The application of these technologies to industrial biotechnology is already well past academic exploration and into commercialization.

Amyris Biotechnologies is pursuing microbial production of biodiesel and a general aviation fuel comparable to Jet-A. The company suggests these fuels will be competitive with fossil fuels at prices as low as $45 a barrel by 2011.[60] Achieving this goal could open up a 3.2 billion-gallon-per-year market—the U.S. Air Force is planning to replace at least half its petroleum-derived JP-8 with synthetic fuels by 2010.[61] Amyris reports that its biofuels have physical characteristics (e.g., energy density, vapor pressure, cloud point) that match or exceed those of fuels derived from petroleum.[62]

In this already competitive sector, the specifics of product development and feedstock requirements tend to be closely held. Yet the number of participants planning to produce fuels in bacteria, yeast, and algae suggests something interesting about the start-up costs associated with entering what is now a highly competitive business, dominated by multibillion-dollar, multinational corporations. Biology offers the possibility of producing fuels at volume, potentially without the substantial infrastructure costs that characterize the traditional petroleum industry. We can expect a profusion of new projects, and companies, along these lines. Assuming these companies are successful, it is worth considering the resulting impact on the liquid fuels market, and more generally the effects on the structure of the economy as a whole.

The economic considerations of scaling up direct microbial production of biofuels are fundamentally and radically different from those of traditional petroleum production and refining. The costs associated with finding a new oil field and bringing it into full production are considerable but are so variable, depending on location, quality, and local government stability, that they are a poor metric of average required investment. But a very straightforward measure of the cost of increasing gasoline and diesel supplies is the fractional cost of adding refining capacity, plus the five or so years it takes for construction and tuning the facility for maximum throughput. Shell recently announced a $7 billion investment to more than double the capacity of its existing refinery in Port Arthur, Texas.[63] The expansion will add 325,000 barrels per day (bpd) to the existing 285,000-bpd refining capacity of the facility, or about 53 million liters per day at a capital cost of about $130 per liter per day. This sum will of course be amortized over the lifetime of the plant, which is probably decades.

This is actually a modest sum compared to the investment required to produce ethanol using *existing* technologies. Facilities designed to produce ethanol by traditional fermentation and distillation, such as the new plant announced by BP and DuPont, can cost as much as $400 million for 420-million-liter annual production capacity, or a bit over 1 million liters per day.[64] This puts the capital cost per liter per day at about $400, about three times as much as the Shell refinery expansion. Paying for this difference in capital cost over the lifetime of the plant will require that ethanol production costs remain low while ethanol prices remain high. Yet if the demand for grain remains high, and if ethanol supply continues to exceed actual market demand, it is not clear that biofuels requiring such significant production infrastructure can easily compete with petroleum products. Moving to biofuels that require less energy input and physical infrastructure could dramatically change the economics of fuel production.

In contrast to the large capital costs inherent in traditional ethanol fermentation, the incremental cost of doubling direct microbial production of future biofuels may be more akin to that incurred in setting up a brewery or, in the worst case, a pharmaceutical-grade cell culture facility. This puts the cost between $10,000 and $100 million, depending on size and ultimate complexity. Pinning down the exact future cost of a microbial biofuel-production facility is presently an exercise in educated speculation. But, for both physical and economic reasons, costs are more likely to be on the low end of the range suggested above.

Butanol, for example, is likely to emerge in the near term as an alternative to ethanol for both physical and economic reasons. Butanol requires less energy than ethanol to separate from water. At concentrations above about 8 percent, butanol in fact begins to spontaneously separate from water, with additional amounts collecting above the aqueous phase. While this concentration is toxic to naturally occurring *Clostridium* species, there is work ongoing to build (or evolve) an organism that can survive under these conditions. Pursuing this avenue of research, James Liao and his colleagues at UCLA have constructed a butanol-synthesis pathway in *E. coli* that operates at 80 percent of the theoretical maximum for organisms grown on sugar.[65] This organism could eventually enable production by fermentation in which butanol is simply pumped or skimmed off the top of the tank in a continuous process. The resulting reduction in energy consumption would constitute an enormous cost improvement.

Costs will fall even further as production eventually moves from alcohols to hydrocarbon biofuels, similar to gasoline, that are completely immiscible in water. Microbes that produce hydrocarbons that float on top of water will dramatically reduce finishing costs for fuel production. The scale required for distillation will no longer serve either as an economic advantage to those possessing such infrastructure or as a barrier to those hoping to enter the market. Moreover, beer brewing presently occurs at scales from garages bottling of a few liters at a time to commercial operations running fermenters that process many millions of liters per year. Thus, once in possession of the relevant strain of microbe, increasing production of a biofuel may well be feasible at many scales, thereby potentially matching supply closely to changes in demand. Because of this flexibility, there is no obvious lower bound on the scale at which bioproduction is technically and economically viable.

The scalability of microbial production of biofuels depends in part on which materials are used as feedstocks, where those materials come from, and how they are delivered to the site of production. Petroleum products are a primary feedstock of today's economy, both as a raw material for fabrication and for the energy they contain. Bioproduction has the potential to displace many of the uses of petroleum and to provide fuel and materials from a very broad range of feedstocks, including the cellulose and other sources of yard waste or sewage available at the municipal level. Eventually, biological production pathways will be enhanced by the addition of photosynthetic pathways, providing a solar boost to the recycling process. Conversion of municipal waste to liquid biofuels would provide a valuable and important commodity in areas of dense human population, exactly where it is needed most. Thus microbial production of biofuels could very well be the first recognizable implementation of distributed biological manufacturing.[66] Someday soon, there is a very real possibility of fueling up your car with biofuels produced within your neighborhood.

Distributed Biological Manufacturing

While transportation fuels are an early target for commercialization of synthetic biology and metabolic engineering, it will eventually be possible to treat biomass or waste material as feedstocks for microbes producing more than just fuels. Dupont and Genencor have constructed an organism that turns starch into propanediol, which is then polymerized in an industrial process into a fiber called Sorona, now successfully competing in the market against petroleum-based plastics. Sorona's competitive advantage comes from building biology into the production process, resulting in an integrated system that is approximately a factor of 2 more efficient than the industrial process it replaces, while consuming considerably less energy and resulting in lower greenhouse gas emissions.[67]

This is just the first step in implementing biological manufacturing, and it is important to highlight the contrast of current technology with technologies to come. Increasingly, the total production of economically important compounds will be miniaturized within biological systems, internalized within single-celled (and, eventually, multicelled) organisms. There is substantial funding behind these efforts.

To much fanfare, BP recently invested $500 million in the Energy Biosciences Institute (EBI), a ten-year project to develop new biofuel technologies at the University of California, Berkeley, the University of Illinois at Urbana-Champaign, and the Lawrence Berkeley National Laboratory. A significant fraction of the funds pledged to the EBI are likely to be used for building new fuel-production and fuel-processing pathways in modified organisms.

Closer to commercialization, Amyris Biotechnologies and Allylix are both working on the implementation of microbial synthesis of isoprenoids, a broad class of compounds with myriad industrial and health-care uses. As a measure of the complexity of what is now possible, recall Amyris's participation in the production of artemisinin in yeast. As of spring 2007, the company and its academic collaborators had demonstrated a billion-fold improvement in yield in about six years (see figure 8.1); it would be difficult to find a comparable historical example of yield improvement in human industrial processes during the last two hundred years. This is just a hint of the potential for biological production as more parts are included in synthetic systems.

Economic success is not guaranteed to the companies discussed above. Markets, influenced by the choices of policy makers, will be the ultimate test of new biological technologies and products. As explored in Chapter 5, even in the context of exponentially improving productivity and exponentially falling cost, decades can pass before new technologies achieve significant market penetration.

But biological technologies will advance even amid economic failure. Over the decades to come, successful implementation of distributed microbial manufacturing could change the economics of production. To be sure, the question of appropriate scale for biofuel production remains undetermined. This is in part due to the early stage of many production technologies, whether industrial or biological. Yet we should be wary of inherited assumptions that the future of biological production will look like historical industrial production. Broadly distributed production using locally available feedstocks could fundamentally alter the way we think about logistics within our economy.

It is not yet clear where, along the range of possible scales, the greatest competitive advantage can be found. It is probably the case that rather than one "winner," the future will be characterized by a greater variety of production schemes, both biological and industrial.

The choice of feedstock—its availability and energy content in particular—will be critical in understanding the extent to which microbial biofuels exemplify the idea of distributed biological production. For example, if the feedstock is primarily sugar from cane, the economics of the operation are probably dictated by the costs of acquiring and shipping the feedstock. Given shipping costs, this might mean production of fuel should take place as close to feedstock production as possible (e.g., Brazil), thereby ensuring that only the most valuable product is shipped long distances. However, if starch or cellulose is used as a feedstock, production could take place much closer to the site of consumption, theoretically in the garage. That is, because starch and cellulose crops can be grown economically in more places than sugar crops can be, technology that enables use of these feedstocks will also enable distributed manufacturing wherever those crops can be harvested or easily shipped.

"The Car Is the Refinery"

The shipping infrastructure for moving starch and cellulose (and sugar) to the consumer already exists all over the developed world in the form of the local market. Processing those feedstocks in the garage would require a domestic production unit. Cars themselves might become a production unit for the fuels they consume. A paper published in the spring of 2007 reported the successful construction of a synthetic pathway, consisting of thirteen enzymes, that turns starch directly into hydrogen.[68] This suggests a future fueling infrastructure in which sugar or starch—substances already available at any grocery store—go into the tank instead of gasoline, ethanol, or any other preprocessed fuel. The hypothetical fueling process is very simple: the consumer adds sugar or starch, the enzymes chew on it, and hydrogen gas bubbles out of the soup and is then used in a fuel cell to provide electric power for the car.

This initial demonstration is a very long way from being a useful fuel source. But it is worth mentioning here because the authors of the report used eleven off-the-shelf enzymes (from spinach, rabbit, *E. coli,* and yeast) ordered from a commercial supplier, and two they purified themselves (from *E. coli,* and the archaea *Pyrococcus furiosus*). Given the obvious utility, it is unlikely that it will be long before those last two enzymes are also available commercially.

It is interesting that this project, in effect, puts on the table new composable parts for producing biofuels (see Chapters 5 and 7), even if the intellectual property is not in the public domain. Moreover, those new parts are reasonably well characterized, at least in vitro. A cursory search of the Registry of Standard Biological Parts does not yet turn up any of these enzymes, but because the gene sequences are either already in public databases or are relatively easy to generate, this gap points to an area of expansion for BioBricks in general and iGEM in particular.

The fact that all these enzymes are as yet unmodified means there is plenty of room for optimizing not just individual enzymes in the pathway but also the whole pathway. The authors note that the highest-yield version of their technology might be a microbe modified to express all thirteen enzymes, in which case a car powered by starch would become something of a cyborg, relying on living organisms to provide power to an inorganic shell.

In response to my description of this technology at a recent meeting of oil industry executives, one attendee demonstrated immediate understanding of the potential, noting, "The car is the refinery." If that innovation comes to pass, a very different marketplace may arise, in which the infrastructure for shipping and refining petroleum overseen by that executive might no longer be such a competitive advantage. Moreover, if distributed enzymatic and microbial processing of simple feedstocks can compete on cost in low-margin commodities markets like liquid transportation fuels, then distributed biological production is likely to make significant inroads in higher-margin markets as well.

An infrastructure for distributed biological production would reduce the value of businesses that presently manufacture and ship goods, and would enhance the value of design services for organisms and genetic circuitry that enable local production. If production fundamentally changes in ways that eliminate the physical advantages of large processing facilities, dramatically different economics come into play. Distributed production will enable participation in markets at different scales, facilitated by enzymes and organisms that "eat" a diverse array of feedstocks. Starch and cellulose will serve important roles in bioproduction as enzymes become available that process these feedstocks economically.

More generally, an increasing diversity of feedstocks and production schemes are likely to support an economic model characterized by more-distributed infrastructure than that of the present economy. It is tempting to speculate that a mature biological economy might rely on a manufacturing infrastructure that displays traits similar to biology itself. In particular, very few organisms on our planet have evolved to be larger than about a meter across. Most biomass production, and therefore most biological processing, occurs at length scales of microns to centimeters. Our future production infrastructure could become indistinguishable from systems of microbes, insects, and cows.

In many cases biological production is already proving to be less expensive and more efficient than traditional industrial approaches for fabricating chemicals and materials.[69] A great deal of investment has recently been made to discover whether the same trend holds true for bioproduction of fuels. But before all of this comes to pass, an economy based on the rational engineering of biological systems is likely to require a library of composable parts with defined behaviors.

Building the Future

Healthy technological economies are not built upon one-off constructions but rather upon hierarchical systems of parts and methods that, when pulled off the shelf and assembled in combination, produce many functions. Economist W. Brian Arthur argues,

> Each component system or assembly of a technology itself has a purpose, an assignment to carry out. If not, it would not be there. . . . And each assembly has its own subassemblies or components. Each of these in turn has an assignment to carry out. Each also is a means—a technology. This pattern, that a technology consists of building blocks that are technologies, that consist of further building blocks that are technologies, repeats down to the fundamental level of individual components. . . . Practically speaking it means that a technology is organized in a loose hierarchy of groupings or combinations of parts and subparts that themselves are technologies. This hierarchy can be as many as five or six layers deep.[70]

Explicitly applying this thinking to the modification of biological systems brings the demonstrated capabilities of rational engineering to a powerful and growing set of composable parts. The biological world we see out the window works in a similar, though not identical, way.

Terrestrial life reuses a nearly universal set of nucleic acids, amino acids, peptide domains, and general classes of whole proteins and metabolic pathways. It appears that historical examples of natural biological innovation—i.e., evolution—often follow genome replication mistakes that introduce repeated genes, in turn followed by reuse and remodeling of the newly redundant proteins and circuits to implement new capabilities.

However, the time scales on which biological systems produce and test new combinations—new designs—are quite different from those required of a technology in human hands. Natural biological innovation is constrained over short time scales by growth and replication of individuals and over long time scales by ecological and geological changes that put pressure on populations. Except in very specialized applications, preexisting biological mechanisms that produce new combinations of parts tend to be slow compared to the lifetime of an individual.

In contrast, biological design and construction by human hands will respond to the demands of the human economy, including safety and security concerns, for which product-development costs are quantified in units of person-days and denominated in the local currency. The product cycle for computing and communications equipment is about eighteen months today, and for consumer electronics only about six months. New design and manufacturing cycles are often scheduled to start as soon as an initial product run is loaded into crates for shipping to market, creating an inexorable march of product model numbers. There is ever-greater pressure to churn out faster, more-capable widgets and gizmos, while margins are constantly being squeezed both by low-cost competition and by ever-decreasing costs for the basic technology.

It remains difficult to apply these notions to biological engineering. Many successful projects to date come from academic laboratories, where costs are often externalized as part of existing infrastructure and where labor is artificially cheap. Yet even in the context of relatively few products with discernable dollar value, whether in biofuels, plastics, or pharmaceuticals, there is clearly considerable progress to note. Measured in many ways, innovation in the development and use of biological technologies is accelerating.

Pushing Innovation Forward

An ability to rapidly innovate within commercial contexts is just one component of the bioeconomy. Regulatory and funding environments are crucial components of the system, with government setting priorities by its preferences. But the substantial funding supplied by governments should be put in perspective.

While significant government support was crucial to the eventual success of aviation and of desktop computers, in both cases commercialization was driven in large part by innovators operating literally in garages. In the early years of aviation, many critical components—primarily control systems, power plant design, and construction techniques—were developed before government funding for the industry materialized in the United States, funding which in any event was originally available in the form of procurement contracts for functioning aircraft rather than as support for research. The development of the technology underlying personal computers received government support for software development, integrated circuit manufacturing and design, and display technology, but it took innovation by individuals in start-up companies to successfully design, assemble, and demonstrate a market for the precursors to today's powerful machines.

In Chapters 9 and 10 of this book, I argued that regulatory regimes that restrict access by entrepreneurs and small firms to engineering tools will cripple the innovation required to maintain economic and physical security. This conclusion is only strengthened by considerations of the larger economy and of the growing impact of biological technologies. The notion that innovation in biological engineering can be maintained in spite of limitations on access to tools and skills is based on a misunderstanding of the history of innovation in the twentieth century.

Whether at the hands of Michael Dell, Steve Jobs and Steve Wozniak, the Wright Brothers, Otto Lilienthal, William Boeing, or the yet-to-be-named transformative individuals working in biology, successful innovation requires wide access to both technology and a multitude of parts. Innovation requires, in effect, a healthy ecosystem consisting of people, ideas, and a great many more pieces than those provided by individual innovators. In other words, innovation requires an existing set of ideas or things that provide the context for the last piece to fall into; that final piece is just one of many. Moreover, as Scott Berkun makes clear in *The Myths of Innovation,* for any given invention or particular scientific advance to have an impact, it has to get it into the marketplace and into the hands of people who will use it: "While there is a lot to be said for raising bars and pushing envelopes, breakthroughs happen for societies when innovations diffuse."[71] Diffusion of an innovation through a society relies on a complex set of interactions that play out at the intersection of the actual technical advance and factors much larger than any individual or group providing that innovation. To tie this argument back to the Technology Line presented in Chapter 5 and to the potential consequences of implementing rational biological design, I will again quote my colleague James Newcomb at Bio-era: "Systems that foster prolific innovation through combination are not just technical; other preconditions include economic, social, and regulatory frameworks that determine the appropriability of value by innovators and intellectual property systems that can support the creative accumulation of innovations over time."[72]

Without a mechanism that enables innovators to capture value from their work—and from the inventions that result from that work—innovation stalls. The ability to build one innovation atop another creates a cascade of new technologies that generates both wealth and jobs. In principle, the mechanism for capturing value from inventions is embodied in patent systems. Unfortunately, as the next chapter illustrates, the patent system as it presently exists often serves to inhibit innovation.

The problems inherent in the present system revolve around a deceptively simple question: What does it mean to own an idea?

CHAPTER TWELVE

Of Straitjackets and Springboards
for Innovation

OWNING IDEAS IS NOWHERE GUARANTEED by the U.S. Constitution.[1] Rather, according to Article 1, Section 8, "Congress shall have
power . . . to promote the progress of science and useful arts, by securing
for limited times to authors and inventors the exclusive right to their respective writings and discoveries."[2] Participating in the "patent bargain"
gives inventors a temporary monopoly in exchange for full disclosure of an
invention and all knowledge and skills necessary to reproduce it. Ideally,
the patent system exists as a mechanism to encourage innovators. In practice, over the last several centuries, it has often been skillfully wielded to
constrain innovation. We will shortly see how patents have slowed the
progress of various technologies, including automobiles and airplanes.

This chapter explores the consequences of patents that may retard the
development of biological technologies. For example, how will we respond
if, in the event of a pandemic, countermeasures are unavailable because development has been slowed not by technical difficulties but by walls set up
by institutions "owning" specific intellectual property? We may soon need
to consider a different means to encourage innovation in biological technologies.

The first U.S. patent law was enacted in 1790, granting inventors "sole
and exclusive rights" to their inventions for fourteen years.[3] That act was
replaced in 1793 by a law that established the Patent Office, with patents
to be granted by the secretary of state.[4] Many revisions have been enacted

since, along with more than two centuries of case law (from judges) and administrative regulations (from the Patent Office); the resulting body of law requires the training of a patent attorney to understand. This is, of course, one reason why patents cost so much to obtain. The initial filing fee in the United States is relatively modest, amounting to a few thousand dollars, but the combination of attorney's fees and filing expenditures rapidly mount to several tens of thousands of dollars. Foreign filing costs can put the total well above $100,000. Thus participating in the patent bargain as an inventor entails substantial costs in addition to whatever is invested in the actual process of invention.

So what does an inventor receive a result of this overall investment? Presently, accumulated (perhaps "accreted" is a better word) law treats a patent as far more than "an exclusive right" to practice an invention or sell copies of an artistic creation for profit. The intellectual content of a patent or copyright is now treated as property to be traded like land or physical objects. Lawrence Lessig, a professor at Harvard Law School and the U.S. Government's "special master" in the Microsoft antitrust trial, argues that the property right now attributed to patent holders, who today are often not the inventors, is not consistent with the original intent of the U.S. Constitution to provide patents as a social good: "Patents are no more (and no less) 'property' than welfare is property. Granting patents may make sense, as providing welfare certainly makes sense. But the idea that there is a right to a patent is just absurd. From the very beginning, our tradition has self-consciously understood that the only question guiding whether to issue a patent or not is whether or not patents will do society any good. . . . Patents are not evil per se; they are evil only if they do no social good. They do no social good if they benefit certain companies at the expense of innovation generally."[5]

According to Lessig, the gradual cementing of control of ideas in the hands of patent and copyright holders is cause for great concern. His book *The Future of Ideas* is ostensibly about the history and future of the Internet. But it is really a book about innovation and what happens to innovation when common resources become privatized:

> The consequence . . . is a change in the environment within which innovation occurs. That change has been, or threatens to be, a shift from a world where the commons dominates to a world where control has been reclaimed. The shift is away from the open resources that defined the early Internet to a world where a smaller number get to control how resources in this space are deployed.
>
> This change will have consequences. It will entrench the old against the new. It will centralize and commercialize the kinds of creativity that the Net

permits. And it will stifle creativity that is outside the picture of the world that those with control prefer.[6]

These words are bold, provocative, and perhaps either offensive or inspiring, depending on the reader's point of view. Lessig makes many assertions about the deleterious effect of intellectual property on the ability to create and innovate, and these assertions may well apply very broadly to technology, including biology. But that is an argument to be made, not accepted outright. It helps to start with a historical example in which intellectual property rights demonstrably, and detrimentally, reduced the pace of innovation.

Flights of Patent Fancy

For insight into the economic future of biotech, it is instructive to examine the transition within aviation from a flight of fancy by a couple of bicycle mechanics to a flight of patent attorneys negotiating the future of an industry.

In his first correspondence with Octave Chanute, dated 13 May 1900, Wilbur Wright wrote, "I make no secret of my plans for the reason that I believe no financial profit will accrue to the inventor of the flying machine, and that only those who are willing to give as well as to receive suggestions can hope to link their names with the honor of its discovery."[7] Chanute responded a few days later: "I am quite sympathetic with your proposal to experiment; especially as I believe like yourself that no financial profit is to be expected from such investigations for a long time."[8] Yet as soon as financial gain appeared possible, various parties began staking claims. The Wrights filed their first patent in 1903, which was granted in 1906. Chanute himself was by no means opposed to the idea of patents in general, having already received several of his own. However, in the case of aviation it appears he considered patents primarily a way of attributing credit rather than a means to financial gain.

In 1908 the Wrights provided design details of their airplane to the Aerial Experiment Association (AEA), a group founded by Alexander Graham Bell. The brothers later discovered that a member of the AEA, Glenn Curtiss, was profiting from aircraft built using that design information, which was protected by the 1906 patent.[9] The Wrights filed a patent infringement suit in 1909 that kept their company, the Wright-Martin Corporation, locked in legal battles with the Herring-Curtiss Company (whose cofounder was former Chanute associate Augustus Herring) for nearly a decade.

In 1910 Chanute gave a newspaper interview that Wilbur Wright interpreted as asserting the Wrights' first patent made no contribution to the state of the art.[10] Wilbur Wright passionately responded with a letter explaining that not only were the brothers' contributions to construction and control methods completely novel but that the Wrights felt fully justified in seeking compensation for their efforts, for which they had paid monetarily and with the sweat of their labor. Moreover, he wrote, the brothers wanted to maintain funds for continued research.

The costs associated with patent litigation, and a reluctance to license, slowed aviation innovation in the United States during the early decades of the twentieth century.[11] Resuming rapid innovation in aviation was of great interest to the U.S. government, as Clarke et al. note, "because the two major patent holders, the Wright Company and the Curtiss Company, had effectively blocked the building of any new airplanes, which were desperately needed as the United States was entering World War I."[12] In 1917, at the encouragement of a committee established by Assistant Secretary of the Navy Franklin D. Roosevelt, U.S. aircraft manufacturers established a patent pool to end litigation, improve licensing, and accelerate innovation. In effect, at the "request" of the federal government, the parties voluntarily relinquished their exclusive rights in order to foster innovation.

Everybody into the Pool?

A patent pool is a contractual arrangement agreed to among patent holders. Depending on the specific terms, the pool may reduce or eliminate royalty claims while providing access to all claims covered by all patents in the pool. As described by University of California, Berkeley, law professor Robert Merges, the patent pool provides a way for parties with both competing and complementary patent claims to cooperate.[13]

Similar agreements have been implemented in many other industries, facilitated by patent property rights. Merges argues that in the automobile industry, in particular, "without the property rights—backed by the threat of production-choking injunctions—the advantages conveyed by the pool would never have been realized. And note: those advantages extended far beyond a cessation of patent hostilities. They included the institutionalized exchange of all manner of unpatented technical information, and the creation of a framework for the crucial task of standardizing sizes and configurations for car parts."[14] To reiterate the central point: Merges concludes that property rights enabled the construction of contractual arrangements through a patent pool. The patent pool, in turn, enabled substantial industrial growth by opening up use and thereby fertilizing innovation in standardized parts. A major reason that some sort of sharing became necessary is that the engineering effort to build new airplanes and cars required utilizing many inventions covered by many different patents. The abundance of patents and their colliding claims created barriers to innovation, in the form of a "patent thicket." This appears to be a general problem, affecting most industries, and should be expected in the various sectors that employ biological technologies.

A profusion of patents in an industry often creates a "tragedy of the anticommons," a concept first developed by Michael Heller at the University of Michigan Law School. Heller and Rebecca Eisenberg applied the idea to biotechnology in 1998: "A resource is prone to underuse in a 'tragedy of the anticommons' when multiple owners each have a right to exclude others from a scarce resource and no one has an effective privilege of use. The tragedy of the anticommons refers to the [complex] obstacles that arise when a user needs access to multiple patented inputs to create a single useful product. Each upstream patent allows its owner to set up another tollbooth on the road to product development, adding to the cost and slowing the pace of downstream biomedical innovation."[15] Merges notes that "once an anticommons emerges, collecting rights into usable private property is often brutal and slow."[16]

There is ample of cause for concern that patents will slow innovation in biotechnology. In industrial biotechnology not only is the number of patents growing rapidly, but so is the number granted *each year*.[17] The backlog of pending applications appears also to be growing. Because many pending patents are not made public for at least a year after the initial application, as the number of pending applications grows, so does the uncertainty about what is covered by the claims included in those applications.

As things stand, any competing or contested claims embedded in these patents, pending or issued, will be sorted out at a later date by negotiation or in the courts. This creates a situation that will inevitably start to resemble the mess that led to the creation of patent pools in the aviation and automobile industries. The U.S. Patent and Trademark Office itself is aware of these issues and in 2000 commissioned a report to examine the utility and feasibility of patent pools in biotechnology. The report found that a patent pool could place "all the necessary tools to practice a certain technology in one place, enabling 'one-stop shopping,' rather than [requiring] licenses from each patent owner individually."[18] "The end result is that patent pools, especially in the biotechnology area, can provide for greater innovation, parallel research and development, removal of patent bottlenecks, and faster product development."[19]

Thus pools *might* provide a way out of an impending biotechnology patent mess. But patent pools will not necessarily be easily implemented for biological technologies, in large part because the field itself is still so new and the relevant case law is still being developed. Moreover, there are larger structural issues that might inhibit creation of, or participation in, patent pools. Few pools are in operation today because they can be used for purposes opposite those intended, leading to government intervention and substantial fines; antitrust action led to the dismantling of the aircraft patent pool in 1975 precisely because the pool was being used by its members to exclude competition.[20]

The Impact of Patents on Synthetic Biology

Amid the discussion on how to proceed with patents in traditional biotechnology, the emergence of synthetic biology is serving to further complicate issues. There are substantial differences between biomedicine as an application area, concerned primarily with marketing individual molecules as drugs, and synthetic biology as a specific approach to biological engineering, a field thus concerned with the use of many tools and pieces of complex systems.

Before I delve further into this topic, it is necessary to clarify what exactly a patent can, and cannot, cover. The U.S. Patent and Trademark Office issues patents on isolated and purified DNA sequences but not on naturally occurring sequences in their original context. The rationale behind the distinction is that isolated or purified DNA, even when derived from a natural source, is the product of human ingenuity. As genetically modified organisms are also the result of human ingenuity, novel genomes and organisms can also be patented. Patents are supposed to provide control solely over artificial constructs and solely over novel contributions, not over information or discoveries. Finally, patents are granted for new functions, whereas copyright protects original works of expression, otherwise known as "content."

In recent years the delineation between patents and copyright has been blurred. Software, for example, is covered by a mix of patent and copyright law, a situation that legal scholars find "far from ideal."[21] Moreover, the scope of patents can be broadened through liberal interpretations by the courts of what defines a patentable invention. Arti Rai and James Boyle, law professors at Duke University, suggest that "there is reason to believe that tendencies in the way the law has handled software on the one hand and biotechnology on the other could come together in a 'perfect

storm' that will impede the potential of [synthetic biology]." The construction of patent pools might serve to resolve some of these issues by both facilitating access to novel technology and alleviating some motivation for overly
broad claims. However, due in large part to the complicated history of intellectual property in both information technology and biotechnology, Rai and
Boyle are pessimistic about the potential for patent pools in synthetic biology
to resolve disputes and maintain access: "Because synthetic biology encom-
passes not only information technology but also biotechnology, the absence
of successful patent pools in the life sciences is cause for concern."[22]

Unsurprisingly, not all interested parties are in agreement about how to
proceed. In particular, just as the Wrights and Curtiss engaged in a battle
that limited innovation and the creation of complex engineered systems,
the existing industry based on biological technologies may be pursuing
policies that limit innovation in synthetic biology. The ability of patent
holders to litigate and restrict the use of patents is a key issue. Presently,
courts often grant permanent injunctions prohibiting the use of patented
materials, which in an age of increasing product complexity can stop innovation in its tracks:

> Because software programs and semiconductor chips are combinations of
> thousands of individual parts, each of which may be subject to an individual
> patent, the risks are considerable that one patent holder, with rights to a small
> component, can enjoin sales of the entire product. Accordingly, the computer
> and software industries have advocated requiring a court to consider the fair-
> ness of an injunction in light of all the facts and the relevant interests of the
> parties. As an alternative to injunctive relief, the court could require only that
> monetary damages be paid. [The Biotechnology Industry Organization] has
> adamantly opposed injunction reform, based on the perspective that most
> biotech products, such a pharmaceuticals, involve only one or relatively few
> patents and therefore face low hold-up risk. Synthetic biology products, how-
> ever, are likely to be highly complex—more akin to software and semiconduc-
> tors—and therefore more vulnerable to injunctive hold-up. One biotech
> company, Affymetrix, which already faces problems that may eventually face
> synthetic biology companies, has separated itself from most other biotech
> companies to advocate limitations on the patent system. Affymetrix produces
> DNA microarrays with thousands or millions of different sequences. Since
> many genetic sequences . . . are patented, a single DNA array might require
> rights to hundreds of individual patents. A single patent holder, with rights to
> one out of thousands of genetics sequences used on the chip, could enjoin
> sales of an entire array. Affymetrix has advocated limitations on the patent
> system with regard to patenting genes or other genetic information.[23]

Large corporations and industrial organizations are lining up not just on
opposite sides of specific patent battles but on opposite sides of efforts to

fundamentally change the patent system. As a result, innovation in biological technologies may depend not only on changes in the interpretation of
existing patent law but also on the outcome of broader debates about the
ownership of intellectual property. These issues are unlikely to be resolved
anytime soon, particularly in light of the considerable financial and political weight of interested parties on both sides of the debate.

The results of the ongoing discussion of the role of patents in fostering,
and hindering, innovation will help determine the course of biological
technologies. Unfortunately, the debate is often predicated upon the notion
that "unlike software development, biotechnology is not likely to be practiced in the garage."[24] The preceding chapters should put the lie to this notion. Garage entrepreneurs and large firms alike must navigate a growing
thicket of patents while struggling to innovate. Even the choice to pursue
patent protection can in fact strongly influence the ability to participate in
a market.

The Era of Garage Biology Is upon Us. Mostly.

My own experience serves as an example. After identifying and document-
ing many of the trends described in this book, it seemed likely to me that
we should be seeing a proliferation of innovation in biological technologies. Many times over the last decade I have written about the possibility—
the probability, and even the inevitability—of "garage biology" and "garage
biotech hacking" based on DNA synthesis.[25] But in about 2005, I realized
I did not know of anyone who was actually operating a business in this
way. At the time, I had an idea I wanted to try, and it seemed like a good
opportunity to test my hypothesis that small-scale, low-cost biotech innovation relying on synthesis, rather than on traditional more labor-intensive
methods, was possible. Based on my previous experience, I was reasonably
certain that the project would be successful in a traditionally funded academic lab or company. But I did not know for certain whether I could pull
it off in my garage. The resulting effort was therefore half start-up company and half art project. You could also call it a bit of experimental economics.

For a very small monetary investment, I found I could make substantial
progress in the garage. However, my labor investment—the sweat equity
of evenings, weekends, and very late nights—vastly exceeded the capital
cost of reagents and tools. When I lacked expertise, or when a particular
piece of equipment I needed was too expensive, it proved possible to hire
contractors or to pay to utilize excess capacity at larger, established com-

panies. For example, I outsourced both synthesizing a gene and expressing and purifying the resulting protein, steps that were inconsequential projects for large labs but that would have cost substantially more if I had to pay for all the infrastructure necessary to build the one little piece I had designed. Then again, this is the way the rest of the economy works. In the end it is thus no surprise, particularly given the arguments I have presented over the course of this book, that garage hacking—that garage *innovation*—has come to biology.

I was careful to choose a relatively simple project to start with—in this case, a tool that should find a use in many different applications—and the circumscribed nature of the project is one reason I was able to keep costs down and make progress in my off hours. By combining parts of proteins described in scientific literature, I was able to specify the design of a new molecule with a particular function.

If this description appears vague to the reader, it is intentionally so, and you have my apologies. For now we come to a serious stumbling block in the commercialization of the products of garage-biology hacking, namely, the problem of how a small firm can participate in a market dominated by much larger and wealthier players. For the time being, securing and defending the property rights so vigorously condemned by Lessig appears to be the best, and perhaps the only, way forward.

One challenge I face is simply the cost of filing a patent. That is, the cost of participation in the market is not just the capital and labor required to produce a new object for sale; it includes the cost of protecting the resulting intellectual property. These costs, and the follow-on costs associated with negotiating relevant contracts and licenses, are often called transaction costs. In my case, the entire capital cost of product development was substantially smaller than the initial transaction costs of writing and filing a patent.

Over the longer term, foreign filing fees and translation costs, maintenance costs, and potentially the costs of defending claims in court will push the overall cost of the patent many times higher. And this is just for one patent, while future filings will certainly add costs. If my company is successful in creating many new tools, or tools with many parts, some of which may be developed by somebody else, then we could wind up trapped in yet another patent thicket. It isn't hard to imagine the costs getting completely out of control. Of course, as does any entrepreneur, I expect that future revenues will be more than enough to cover these and other costs. In any event, innovation in biological technologies has now reached a point where other industries have been for some time: the cost of demonstrating an advance can be less than the transaction cost of protecting the intellectual property that the advance represents.

This observation encapsulates a quandary for me. Participating in the present market *requires* spending considerable sums on a patent. If it were possible to forgo patent protection, then with a modest amount of additional development I could potentially begin selling a tool that provides a new quantitative capability in molecular biology. But by merely describing the functionality of the tool, an obvious necessity from a marketing standpoint, I would enable competitors to reverse-engineer the tool. While it is true that the investment required to innovate in biological technologies has fallen, the industry is still dominated by Goliaths that have the physical and financial capacity to rapidly commercialize new ideas. This is, of course, the reality that many other entrepreneurs must face. Yet in addition to any limitation on innovation from patent thickets and the property staked out by Goliaths, the present structure of intellectual property protection and the associated transaction costs serve without question as a straitjacket on innovators. This straitjacket limits the ability to invest in new technologies by constraining the ability to participate in the market. In order to pay the rent, I am forced to don this particular straitjacket.

The present cost of participating in the market therefore inhibits the development of the very division of labor that William Baumol suggests is the optimal arrangement to produce maximal innovation, as presented in Chapter 10.

It would seem that a burgeoning biotech economy, built from below via invention and innovation by entrepreneurs and small firms, would benefit by a structure that enabled less capital-intensive protection of ideas. From another perspective, investments in new technology would provide a higher rate of return if less initial capital were required to participate in the market. Creating a mechanism that allows the Davids to participate in a biotech economy already dominated by Goliaths, perhaps by lowering transaction costs, is crucial to enabling rapid innovation in biological technologies.

The Trouble(s) with Patents

Patents are supposed to benefit the public good by providing innovators a temporary monopoly on their invention, in exchange for disclosing that invention so that others may learn from it. Because granting a property right on an invention is explicitly an economic exchange, which is today often seen as a lever generating significant financial benefit, there is motive for those innovators with sufficient resources to abuse the public trust and attempt to create a monopoly far larger than what may be justified by the substance of their patent filing.

Many granted patents do not in fact meet the standards set out by Congress and the U.S. Patent Office. A recent study of patents on human genes associated with a certain set of diseases found that a large number of those patents "do not measure up to the federal patent law." The particular diseases followed in the study were chosen because they are in the public eye and they represent an identifiable area in which problematic gene patents "could potentially have an impact on human health care." The study examined seventy-four patents, filed over a one-year period, which in total contained 1,167 claims. Almost 40 percent of those claims were problematic, with some containing more than one problem, with "problem" defined as insufficient written description and enablement; claims for "far more than what the inventor actually discovered"; claims of discoveries that the "patent holder did not specifically describe"; claims of utility of an invention not actually demonstrated in the patent; and claims of utility based on correlation between the presence of mutation and disease without demonstrating how that correlation could be used to diagnose the disease.[26]

The patent system is frequently exploited in this way. Many patent applications attempt to claim extraordinary protection of ideas and technological capability, in particular by claiming in patents ideas that are common, though unwritten, knowledge in a field. Yet worse for small firms, because of the cost of participating in the process, patents are tools requiring ever more resources to wield. This puts the advantage squarely in favor of larger businesses. Moreover, the system requires that smaller firms and entrepreneurs find financial backing to cover transaction costs that may be substantially in excess of the funds required for invention and initial innovation. Given the political reality of influence granted to corporations by dint of their relative wealth, I do not expect that this situation is going to change radically anytime soon.

Patents may simply be a cost—and a costly way—of doing business. However, there are other ways to protect intellectual property and, indeed, to organize economies and markets. It is worth exploring whether these alternatives could facilitate innovation in biological technologies by providing "rights" to innovators while lowering transaction costs.

Opening Up Innovation

Octave Chanute pointed the way very early on (see Chapters 3 and 5). With regard to competing suggestions for building airplane control systems, in particular those that would enable "automatic equilibrium" or "obtaining automatic stability in the wind," Chanute followed his habit of trying to maximize the flow of information in order to speed innovation. In 1910 Chanute suggested to Charles Walcott, secretary of the Smithsonian, a technology-testing program that would result in a standard platform for future innovation: "It would be my idea to gather information upon these various proposals [of Langley, the Wrights, Hargrave, Montgomery, etc.], to investigate them and to test the more promising . . . and then to publish the results for the benefit of aviators who want to apply motors."[27]

Today this strategy might be called "open-source aviation," after the cultural and economic phenomenon known as "open-source software." Alternatively, the strategy might instead fall into the category of "open innovation," a means of collaborative innovation in which many users contribute to the development of product. Open source and open innovation are distinct concepts—concepts that are too often conflated—that I will discuss in the remaining chapters of this book.

Of course, it is neither fair nor accurate to retrospectively put Chanute's efforts in this context. Open source and open innovation are creations of the late twentieth century, enabled by new modes of communication and new realities of highly competitive global business. But the seed of the idea was clearly there—Chanute understood that sharing information would more rapidly lead to technical understanding that could then serve as a basis for additional innovation.

There is substantial and growing support for open innovation among the largest corporations on the planet. The reasons are simple: the design process for many products has become so complex, and the flow of information so rapid and inexpensive, that collaborating has a higher value in the market than does hiding behind walls.

IBM—Big Blue—was once among the most successful manufacturing organizations in the world. During most of its history, IBM's innovations were largely derived from internal development efforts. Now the company allows open access to its most sophisticated supercomputers, and it markets open-source software, obtained at no cost to IBM, written largely by programmers working for free.

Why the change, and what enabled this strange new economic structure? Harvard law professor Yochai Benkler provides one answer: "Because [IBM doesn't] see it as a choice. They understand the world is becoming too fast, too complex and too networked for any company to have all the answers inside."[28] IBM's hardware sales are in part supported by the widespread use of Linux, and IBM itself supports Linux by funding development in areas it finds useful. Thus, while IBM appropriates considerable value provided by volunteers, it also contributes to the maintenance and development of code that those volunteers use.

How does IBM generate revenue with this model? The money is in building and maintaining computing equipment and in providing services rather than simply in producing and selling goods, as explicitly acknowledged by IBM CEO Lou Gerstner: "Building infrastructure—that's our franchise."[29] Benkler notes that while "the firm amassed the largest number of patents every year from 1993 to 2004," between 2000 and 2003 "the Linux-related services category moved from accounting for practically no revenues, to providing double the revenues from all patent-related sources, of the firm that has been the most patent-productive in the United States."[30]

Despite this impressive showing, IBM's strategy should probably be considered an experiment by a company dealing with an economy-wide shift in the United States away from manufacturing. IBM has implemented a fundamental change from a business model based on selling objects to one based substantially on selling ideas and services, including consulting services related to the Linux packages it distributes for free. In 2006, of IBM's revenues of $91.4 billion, about 53 percent was derived from services, 25 percent from hardware sales, 20 percent from software sales, with the remainder from finance operations.[31]

Kenneth Morse, head of the MIT Entrepreneurship Center, has a more-cynical view of IBM's behavior: "They're open only in markets, like software, where they have fallen behind. In the hardware market, where they have the lead, they are extremely closed."[32] The company continues to pursue patents in areas where that strategy can provide long-term value, such as advanced materials.[33] Still, at least for the time being, IBM appears to be profiting handsomely from its selective turn to opening up innovation.

IBM's strategy is one version of "open innovation," a concept championed by UC Berkeley professor Henry Chesbrough that has become something of a business mantra in the last few years.[34] Chesbrough first defines the mind-set of "closed innovation," which considers that "successful innovation requires control"—control of ideas, design, manufacturing, sales, service, finance, and support.[35] Chesbrough asserts that this model is no longer viable because of a combination of factors, including the mobility of skilled labor, the availability of capital to finance innovative competitors, a reduction in both time to market and longevity of new products, increasingly knowledgeable customers and suppliers, and international competition.[36] "Open innovation," in contrast, "is a paradigm that assumes firms can and should use external as well as internal ideas, and external and internal paths to market, as the firms look to advance their technology."[37]

One large firm embracing these ideas is Procter & Gamble (P&G). Over the course of the last ten years, P&G has increasingly looked outside the firm, ramping up the proportion of products derived from outside ideas from less than one-fifth to more than one-half. Through this version of open innovation, the firm grew by 6 percent a year, tripling annual profits to $8.6 billion, and reduced the failure rate of new products from 80 to less than 50 percent.[38]

Chesbrough's book, *Open Innovation: The New Imperative for Creating and Profiting from Technology,* catalogs many such examples of greater information flow into and out of companies leading to greater revenue.[39] Another variant of these ideas is "user-driven innovation," chronicled by Eric von Hipple, a professor at MIT's Sloan School of Management. User-driven innovation describes an explicit conversation with customers not only to produce a product they prefer but also to facilitate customization: "Networks of hyper-critical users can help firms quickly filter out bad ideas, and thus encourage fast failing."[40] Which brings us back to Chanute and his suggestion of pursuing more-rapid innovation through collaboration.

Yet this all begins to sound less like a fundamental revolution in innovation and more like better salesmanship, or perhaps a better business model. New business models constitute innovations, to be sure, but they primarily serve to enhance competitive abilities through improved productivity or customer service, while usually maintaining the traditional structural relationship between the customer and the producer. I certainly do not mean to discount the value of new business models in helping foster innovation in biological technologies, and Von Hipple's user-driven innovation is part of a larger phenomenon of democratizing innovation, which is an explicit theme of this book. But none of the discussion above serves to truly address the fact that the "exclusive right" presently construed in patent law is likely to impede innovation in biological technologies. Open-source software, in contrast, is an example of a fundamental shift in the structure of production, enabled by a clever reimagining of the "exclusive right," a reimagining that will teach us more about what is required to avoid stagnation in biology.

An "Amazing Phenomenon"

Open-source software (OSS) presents substantial challenges to traditional economic and organizational models of production. In a seminal paper on the phenomenon of open source, Benkler wrote,

> At the heart of the economic engine of the world's most advanced economies, and in particular that of the United States, we are beginning to take notice of

a hardy, persistent, and quite amazing phenomenon. A new model of production has taken root, one that should not be there, at least according to our most widely held beliefs about economic behavior. The intuitions of the late twentieth-century American resist the idea that thousands of volunteers could collaborate on a complex economic project. It certainly should not be that these volunteers will beat the largest and best-financed business enterprises in the world at their own game. And yet, this is precisely what is happening in the software industry.[41]

Benkler proposes that a new form of production is emerging in the networked world, "commons-based peer production," of which open-source software is one example. "The central distinguishing characteristic" of this new form of production "is that groups of individuals successfully collaborate on large-scale projects following a diverse cluster of motivational drives and social signals rather than market prices or managerial commands."[42]

The "motivational drives" behind the open-source software movement in particular have both practical and ideological origins, about which I will say very little here. It is not my goal to retell the story of the open-source movement, to examine the personalities involved, or to explore the historical contingencies that led to the creation of particular software packages or social organizations. Nor is it my purpose to explore all the ways that open source can be used to generate financial profit from software. Rather, my purposes are to explore (1) what makes OSS, and peer-production in general, a source of innovation and (2) whether the model is transferable to the development of biological technologies and the bioeconomy. As an entrepreneur in particular, and as someone concerned about the pace of innovation more generally, I want to try to understand whether there are any lessons to be learned from OSS that can help me develop biological technologies both quickly and safely.

The feature of OSS most relevant to this book, and to the future of biological technologies, is that it is far more than a description of a particular software product, a particular package of programs, or even simply a set of licenses that governs the transfer of copyrighted code. Open source is a fundamentally new model of production, one that for the moment relies very heavily on a particular interpretation of the rights of the people who write and use code.

The Legal Basis of Open-Source Software
as a Commons

As a practical matter, the phenomenon of OSS, as presently constituted in the world, rests in large part on the existence of Linux and BSD, two versions of the UNIX operating system that can be used for free and modified at will. If the substance of the open-source phenomenon is the use and improvement of a commons comprised of code in Linux and BSD, the rules governing—in fact, requiring—participation in the commons are nearly as important as the code itself. As described by Steve Weber, a professor of political science at UC Berkeley, the intellectual property regime that supports OSS "takes shape in a set of "licenses" written for the most part in the language of standard legal documents. . . . Think of these licenses as making up a social structure for the open source process. In the absence of a corporate organization to act as an ordering device, licensing schemes are, in fact, the major formal social structure surrounding open source."[43]

Very briefly, open-source licenses are built on the copyright of code. The original holder of the copyright, namely, the person who wrote the original code, determines the terms of use by releasing that code under a given license. Without a property right, in this case conferred by copyright, contracts governing the use and distribution of code would have to be arranged on a case-by-case basis. As it stands, when you, the user, acquire and read the source code, you explicitly agree to the license. The licenses usually distinguish between use and distribution. If you obtain code under an open-source license, you may generally use it however you like for your own purposes, including changing the code. However, the licenses usually stipulate that if you distribute the altered code in any form, you are contractually obliged to make any altered source code available under the original license. The licenses thereby guarantee an ever-expanding commons of source code, a characteristic that is sometimes called "viral," in that sharing source code leads to further sharing. Using the exclusive right granted by copyright to ensure viral spread of code is also called "copyleft."

The commercial consequences of this strategy are immediately apparent. After investing your labor and financial resources in improving code under a copyleft license, you must give the product away, even if you can claim copyright on a portion of the improved code. Most copyleft licenses do not stipulate that revenues from selling code be shared with the original copyright holder. Thus the fundamental conundrum facing economists and lawyers who specialize in intellectual property: why would anyone contribute to a project for free that others may make money on? Steve Weber addresses the question very succinctly: "Stark economic logic seems to undermine the foundations for Linux and thus make it impossible."[44] One obvious resolution is, of course, to argue that standard economic logic is insufficient to describe the new phenomenon.

The Economic Impact of Open Source

In his book *The Success of Open Source,* Steve Weber takes the very instructive view that "open source is an experiment in building a political economy—that is, a system of sustainable value creation and a set of governance mechanisms. In this case it is a governance system that holds together a community of producers around this counterintuitive notion of property rights as distribution. It is also a political economy that taps into a broad range of human motivations and relies on a creative and evolving set of organizational structures to coordinate behavior."[45]

The "sustainable value creation" of open source is both social and financial in nature, and the experiment appears to be succeeding. The growing software commons is a shared resource that anyone can draw on and contribute to. And the existence of that software enables real wealth creation in the form of jobs and many billions of revenue in software and hardware sales and support. That hardware and software then serves as infrastructure for a large fraction of the economy as a whole.

Anyone who uses Google as a search tool is relying on more than one hundred thousand servers running Linux.[46] Google's search results themselves are an example of Benkler's peer production: the PageRank algorithm determining the order of Google's search results "relies on the uniquely democratic nature of the web by using its vast link structure as an indicator of an individual page's value."[47] Open-source software has broad reach. According to Dirk Hohndel, Intel's Chief Technologist for Open Source, "Of the top 500 fastest super computers in the world, currently half of them are running on Linux."[48] Open source is also maintaining its dominance in more-prosaic service sectors; somewhere between one-half and three-quarters of servers connected to the Web are running open-source software, depending on the month and on which numbers you look at.[49] As far as Big Blue goes, in addition to revenues from Linux-related services, IBM reportedly saves $400 million per year by relying on Linux.[50]

The broader economic impacts are substantial, and we are probably seeing just the tip of the iceberg. A recent report from the European Commission found that replacing the open-source code presently used within European firms with software written in-house would cost approximately $18 billion and that the code base has doubled every eighteen to twenty-four months for the last eight years.[51] The report also found that the share of the economy related to, and supported by, open-source software could reach 4 percent of European GDP by 2010.

This extraordinary economic value is built upon the use by programmers of licenses that explicitly reduce or eliminate transaction costs for all involved. Linux is a common resource that anyone can employ at virtually no cost. The relevant point here, and the reason that many programmers appear to feel comfortable contributing code to the commons, is not that the code is free but that the license in principle gives everyone an equal opportunity to profit (monetarily or otherwise) by contributing, packaging, and distributing code. That said, while some licenses do specify how economic value from open-source code is to be returned to the copyright holder, most of the substantial sums generated by selling and using OSS do not accrue to the people who wrote the code. Moreover, large organizations clearly have an advantage when it comes to supporting the necessary labor and physical infrastructure required to give away something of value, and clearly corporations make plenty of money from a resource created, and curated, by somebody else.

There are many explanations for this phenomenon floated in the literature on open source. One answer may be that programmers accrue social capital—reputation—by contributing code. Another is the simple pleasure gained from contributing code. Yet another is that the development of some open-source code is paid for by corporations, governments, and foundations, which, despite being required to give the code away, then leverage that code to generate other value.

One motivation for the original General Public License (frequently referred to just as "the GPL"), under which a large fraction of OSS is distributed, is the assertion that the person who wrote the code should have the freedom to dispose of it however he or she wishes. Thus "free" open-source software, as Richard Stallman, the president of the Free Software Foundation, says, is "free as in free speech, not as in free beer. Free software is a matter of the users' freedom to run, copy, distribute, study, change and improve the software."[52] Stallman, with the assistance of legal counsel, wrote the original version of the GPL in 1989.[53]

In contrast, Linus Torvalds, who released the original version of what became known as Linux, professes a greater interest in the process: "I think it's much more interesting to see how Open Source actually generates a better process for doing complex technology, than push the 'freedom' angle and push an ideology."[54]

Regardless of the specific motivation, in the face of considerable history and economic might, open-source advocates have asserted their right to do something different. And it works. That open source not just exists but thrives is a testament to both the ingenuity of these innovators and the flexibility of constitutions and laws of countries around the world. This success comes despite the fact that the laws and legal precedents determining the economic value of software are "far from ideal."

Hacking the System

It is then with the greatest respect that I write that the GPL and the structure of the open-source software movement often looks to me like a hack. That is, because of accreted legislation and court rulings—which constitute a bunch of after-market regulations bolted on to preexisting laws in order to accommodate the protection of a technology that none of the original authors of the U.S. Constitution could have foreseen—there is no rational, engineered solution to either protecting software and algorithms as intellectual property or to ensuring that they remain available to all in a commons. Instead, Richard Stallman and others have constructed an extremely clever mechanism for preserving their right—their freedom—to not only do what they wish with the code they write but also to specify that others who use their work share the results. It seems to be a popular hack.

Viewed as socioeconomic technology, open source is rapidly penetrating the market and is being adopted much faster than many other technologies have over the last century (see Chapter 5). The use of copyright and contract law to protect the right of innovators to specify how their code can be used is therefore among the most beautiful, successful, and, thus far, stable hacks in the history of computing.

As anyone who has written code can testify, hacks can become embedded in software, never to be really smoothed out or documented properly, and thereby never to be properly understood or described. Hacks often start as work-arounds, or quick fixes. This does not mean the hack is necessarily bad nor that a complete rewrite of the code is necessary. The hack may suffice: it works. The same phenomenon is familiar to people who produce physical hacks of hardware, homes, cars, and boats.

Of course, the GPL and OSS are perhaps better viewed not as coding hacks but rather as hacks of our socioeconomic system. And these hacks seem to work *really* well, not only providing a springboard for innovation by software developers but also adding substantial and measurable value to the economy as a whole.

This raises for me yet another conundrum, or at least the appearance of one, because the hack works as a result of the very declaration of explicit property rights that so concerns Lawrence Lessig. The resolution of the conundrum is in the assertion of a commons, to which contributions can be easily made. The successful establishment (and defense, so far) of a software commons in no way lessens the importance of Lessig's arguments about the fencing off of public resources as private property. But it also highlights the importance of giving innovators some means of asserting their rights over the product of their labor *on their own terms*. The decision by the U.S. Congress to extend copyright protection to code provides a foundation to construct whatever (legal) contractual obligations one desires. The open-source model now constitutes a middle ground between proprietary technology development and placing knowledge in the public domain, freely revealing it for anyone to use.

The most fascinating aspect of this structure for me is not the difference between open source and proprietary development, which seems to occupy most commentators, but rather the difference between open source and "free revealing." There was never any legal or technical barrier standing in the way of Stallman or Torvalds putting their code in the public domain. That code would have formed part of a commons, which could have been maintained and improved over time. But the implicit copyright on the code would have prevented broader use and would have created a rapidly growing copyright thicket that all developers would have had to work around. In contrast, the GPL provides a means for people to collaborate on maintaining a garden that serves the needs of any who care to participate. The viral, copyleft licensing strategy does not prevent anyone from selling code found in the commons, but it does allow developers to require that, if their work serves as the basis for improvements, any improvements remain accessible to all.

As I see it, there are three factors that contribute to the utility and success of the open-source socioeconomic hack: (1) the explicit declaration of the software commons as a resource to be maintained and enhanced, (2) the creation of licenses that guarantee the rights of the creator while reducing transaction costs and enabling anyone to utilize the commons as either a producer or a consumer, and (3) the minimal capital and labor cost required to create and copy bits. The hack took place within the existing system of socioeconomic expectations about the behavior of people and markets and was itself unexpected and initially hard to explain. But no laws were changed to enable OSS, and no professors of economics, law, or political science, or anyone else for that matter, were put out of business when OSS cropped up. In contrast, there is a remarkable profusion of economic value, jobs, and professorial output that is a direct consequence of OSS.

Bruce Perens, an early participant in the OSS movement, contends that there is in fact no mystery, that open-source code and the attendant licenses are the result of programmers creating and exploiting a new niche in a market economy: "Open Source can be explained entirely within the context of conventional open-market economics. . . . Open Source is self-sustaining, with an economic foundation that operates in a capitalistic manner. It does not require any sort of voodoo economics to explain. It is an extremely beneficial component of a free-market economy, because of the very many people and businesses that it enables to make their own economic contribution."[55] That is, open source works within the market precisely because it provides a way for people to participate on their own terms, even if those arrangements are not obviously understandable in terms of traditional pricing models. Recall the discussion in Chapter 10 that classical economics does not adequately value innovation; open source works precisely because it gives value to the process of innovation and allows innovators to set the terms under which their exclusive right is utilized by others.

Therefore, just as predicted by William Baumol (see Chapter 10), innovative entrepreneurs are the "economical supplier of breakthrough innovation to the economy," in which those breakthroughs are composed of not just code but also the novel socioeconomic structure of open source.[56]

Perens extends his economic explanation to areas beyond software, posing and then answering, the question "If open source works, why don't we all build our own cars?"

The Open-Source paradigm works well for many products where the major value of the product is its design. It's most successfully been used to date to produce software, an encyclopedia, and integrated circuit designs.

The integrated circuit designs are programmed into a field-programmable gate-array, a special device that can be programmed with a design and will then immediately behave as a circuit with that design. This is an example of a field in which hardware is very much like software.

However, most things are not software. It only takes a cent's worth of resources to make a copy of a piece of software, but it takes a pound of flour to make a loaf of bread. Someone has to farm the wheat and grind it into flour, and those efforts have to be paid for.

Automobiles, of course, are much more complex than bread, and it takes a great many physical processes with expensive tooling to manufacture them. Consider that to make an electric motor, one must mine and refine metal, then draw wire, roll sheet metal, cast and machine bearings, and then assemble all of these pieces into very precise forms. It should be no wonder that it takes an entire economy to manufacture an automobile, while a single individual may produce an important software product.

When the day comes that we can make complex physical products by producing their designs and directing a machine to manufacture them from easily-available materials and electricity, the economy will change radically. Today, we are limited to producing individual parts with computer-controlled milling machines, a slow and dirty process that still requires manual intervention. A healthy Open Source community has evolved around such machines, and we are starting to see them share part designs. But the science of computer-controlled manufacturing will have to improve tremendously before we can have "Open-Source cars."[57]

And what of the day when we can make complex biological products by directing a machine to synthesize a newly designed genetic circuit, then use an organism to run that circuit, and thereby produce a new object or material using easily available feedstocks and sunlight? That day, as this book describes, is here. This observation establishes the groundwork for considering to what extent the development of biological technologies can benefit from open source in support of breakthrough innovation by entrepreneurs.

Open source, as described thus far, is enabled in practice by economic factors. It is certainly true that without the voluntary participation of programmers, and the remarkable organizational structures within which they work, the product itself would not exist. But the creation and distribution of the product relies upon low costs for writing and copying code and upon low (or nonexistent) transaction costs to transfer the right to use code.

I first heard the phrase "open-source biology" sometime in the late 1990s from my colleague at the Molecular Sciences Institute, Drew Endy. Inspired by the growing success of OSS, and wary of the effects of broad patents on biological innovation, starting in about 2000 I set out to understand how the idea of open source could be used to accelerate innovation in biology.[58] The next chapter explores what open-source biology might be, who is thinking about it, and what it would take to put it in place.

CHAPTER THIRTEEN

Open-Source Biology, or Open Biology?

BIOLOGICAL TECHNOLOGIES are experiencing exponential change. Garage biology is a reality. India, China, and Brazil are among many rapidly developing economies pushing homegrown biological technologies. Undergraduates and high school students around the world are contributing to high-end research and development. What a world we live in.

We are even seeing the beginning of collaboration motivated by complex engineering. As related in Chapter 11, large corporations seeking to develop multitrait genetic modifications of plants have recently entered cooperative agreements precisely because the job is so complex. The challenge ahead is in facilitating that sort of collaboration more widely, particularly in the face of resistance by many companies and organizations that presently see profit in maximizing proprietary control of individual genes or gene products.

In 2000, motivated both by the specter of a fenced-off set of proprietary parts and methods and by the potential of open-source biology to encourage sharing, I coauthored a white paper suggesting a concrete open-source biological engineering effort. The white paper was communicated to DARPA, the research arm of the U.S. Department of Defense, requesting

> funds to begin to develop and maintain a body of publicly available technology, to foster a community of researchers who contribute to this open-source

technology repository, and to publicize the concept and the actual workings of open-source biology through meetings and the web. Our near term goal is to generate a set of interoperable components sufficient to comprise a basic "kernel" or basic "OS" for phage, bacterial, viral, plant, and animal systems. Components of the core OS will include tissue specific gene regulatory elements, transcription regulatory proteins and sites whose activities are tunable and switchable by small molecule chemicals, site-specific recombinases, and protein domains that can be used to direct specific protein-protein interactions.

Like other distributed systems, biological research and biological engineering efforts conducted in an open source manner will be robust and adaptive, providing for a more secure economy and country.[1]

The proposal was not funded, and for good or bad the U.S. government has made very few investments that might be construed as developing, or even as maintaining, broad and open access to biological technologies.

"BioBrick Parts" as Open Source

The BioBricks Foundation (BBF) was founded by Drew Endy with the following goals:

- to develop and implement legal strategies to ensure that BioBrick standard biological parts remain freely available to the public . . . ;
- to encourage the development of codes of standard practice for the use of BioBrick standard biological parts . . . ; and
- to develop and provide educational and scientific materials to allow the public to use and improve existing BioBrick standard biological parts, and contribute new BioBrick standard biological parts.[2]

The growth in participation and the achievements of those "competing" in iGEM suggest that, as far as maintaining wide access goes, the BBF is on the right track. Yet, as of this writing, at the beginning of 2009, there is no framework in place to ensure that BioBrick parts remain freely available other than to place those parts in the public domain.[3]

The BBF is faced with trying to create a commons based on thousands, and soon tens of thousands, of parts. Filing patents on all these parts to secure their use within a patent pool or other contractual institution would be prohibitively expensive. Moreover, in order to maintain the integrity of the licenses, all the claims in all the patents would require defending, which immediately opens the door to oceans' worth of litigation that would drown the BBF. Thus the BBF, at least initially, is simply putting all the BioBrick parts into the public domain, a strategy that, as Rai and Boyle note, "not only makes the parts unpatentable, but . . . undermines the possibility of patents on trivial improvements."[4] As an organization, the cost to the BBF of accumulating parts is small for the time being—the labor and infrastructure are supported by universities and other nonprofits. Thus the cost of the intellectual property protection vastly exceeds the cost of invention, and the BBF is facing the same sorts of issues that I, as an entrepreneur, am attempting to navigate while innovating on a small scale, one part at a time (see Chapter 12).

The philosophical bent of the BBF might be described as viewing DNA as a close analogy to computer code. Explicit references to "programming DNA" suggest that the engineers involved see designing new circuits and new parts as very much a software problem (see Chapters 7 and 8). As the previous chapter describes, the simplest and lowest-cost means to secure a property right on software is a copyright, which can then be licensed via contract law. There is a clear fracture in the analogy between open-source software and open-source biology at this very point.

There is no legal basis for applying copyright to biological constructions. The observation by Steve Weber that "licensing schemes are . . . the major formal social structure surrounding open source" suggests that the social structure surrounding biology is determined either by public domain exchange of information or by patents.[5] Indeed, the organizations engaged in basic biology or in the development of applications are either explicitly nonprofit (predominantly academic institutions funded by grants) or explicitly for-profit (predominantly large corporations funded by sales). This is a substantial limitation on the freedom of scientists and engineers to self-determine how they participate in the market. Open source is a third way for software developers to participate, but this option is not available to those innovating in biology.

Patents are far from an ideal mechanism for serving as incentive for either a means of developing biological technology or as an organizing principle for social structures. Some other sort of "exclusive right" needs to be invented, perhaps in the form of another socioeconomic hack enabling a different sort of commons, which innovators could use to choose the terms upon which they participate in the market.

Another Definition of Open-Source Biology

Richard Jefferson is a man with a mission. Jefferson is well known as the inventor of several technologies that facilitate the genetic modification of plants. He is also well known for deciding that, rather than charging what he views as exorbitant fees demanded by many biotech companies, the best way of getting his innovations into the hands of people who want to use them is to let users pay what they are able. Monsanto, as a multinational corporation, paid "a substantial amount," while a much smaller firm paid just enough for Jefferson to buy a classic guitar.[6] Researchers in nonprofit or academic organizations get access for free, and small for-profit companies in developing countries might participate for in-kind contributions of labor or other support.[7]

The fees from Jefferson's patents enabled him to found his own research organization, CAMBIA, in Australia and to try out his own version of open source. The effort is organized around patents. Jefferson is quite clear that he is attempting to change a system in which he views patents on biological technologies as tools that are often used by large organizations to create barriers to competition: "When big companies invent, discover, or acquire these technologies they rarely use patents to generate and share the next generation of technology."[8]

Jefferson's initial effort takes the form of a license agreement for biological technologies. A biological open source (BIOS or BiOS) license typically stipulates that, in exchange for the royalty-free right to use a technology, you agree not to assert property rights, including patents, in any way against other parties who agree to the same license.[9] Note that the "royalty-free" grant does not exclude a fee (a "technology support subscription") for permission to use the technology, a fee that enables CAMBIA to pay its bills. The fee schedule listed on the licenses ranges from zero to $150,000, depending on the size of the organization signing the license.

Another important feature of the licensing strategy is the inclusion of a mandatory grant back to CAMBIA. Once you sign up to for a license, if you improve the technology, you must provide a license back to CAMBIA for the improvements. Thus the pool of technologies grows.

BiOS licenses are explicitly structured to provide for nonexclusive commercialization, which creates a permissive patent pool administered by CAMBIA.[10] If you agree to use a patent under a BiOS license and then develop improvements, you must not only allow access to those improvements but also ensure that other members of the pool have exclusive access; the goal of using a BiOS license is the "development of a protected commons around the technology."

Upon improving technology secured via a BiOS license, presumably at some financial cost, as a licensee you appear to have two options. You can give away the improvement, putting it in the public domain and thereby giving up any rights, *including the right to ensure that any further improvements are secured for the pool.* Alternatively, you can attempt to maintain your rights in some fashion, but according to the license terms, you must make improvements available to any other BiOS licensee. A patent is presently the only legal means of maintaining such rights. Implicitly, therefore, you *must* seek patent protection on any improvements in order to secure rights that may then be made available to other members of the pool. In this case, the "source code" being opened up, and constituting the "property" in the commons, is not actually biology but rather the substance of a patent. The "operating system" for BiOS licenses is some complicated combination of patent law, contracts, engineering, product development, and marketing; in other words, the operating system approximates the larger legal system and economy.

To be clear, this style of license defines not a public space but a walled garden curated by CAMBIA, access to which is granted only through the purchase of a key. Even if an organization pays a very small fee for access to the pool initially, giving improvements back to the pool requires pursuing patent protection. BiOS is built to work in the context of existing intellectual property laws. But it is not a garden that just anyone can play in. Even making sure you understand a BiOS license will require hiring an attorney. Moreover, because maintaining the commons itself incurs costs, such as transferring materials and paying for IT infrastructure, CAMBIA asks "for-profit licensees to pay some of these costs, at rates related to size of the enterprise."[11] Through the variable fee schedule, Jefferson is attempting to create terms for participation that are tailored to the organization desiring a license; bigger companies pay more than start-ups and academics. However, the fees accrue to CAMBIA, not the original owners and developers of the technology. They must compete and survive through participation in the market; they must cultivate paying customers. While BiOS constitutes an excellent experiment, it is not yet apparent that the "protected commons" approach typified by BiOS will in fact enable widespread innovation, because it is not clear how small entities can use the commons to turn a profit.

As was discussed in the last chapter, software can be "free" in both senses of the word—economic and personal freedom—because it costs very little to reproduce. The use and distribution of software can be protected via implementing inexpensive copyrights and contracts, and the spread of both the software and its licensing agreements requires very little capital. In contrast, because biological innovations presently require a patent for protection, even an "open source" biological commons implemented via contract law requires first expending significant capital to protect each technology in that commons.

Jefferson's goal is laudable; accelerating innovation is a major theme of this book, and he may well have a good mechanism to help. Unfortunately, the existing landscape of patents and large corporations poses substantial barriers to progress, barriers that slow down even those corporations. Costs in biotech are falling exponentially, and complexity is now being built from the ground up; both trends enable a very different kind of innovation than has existed previously. A very real concern for the future is that so many complex and broad patents have been filed and granted not just on biological products but on basic technologies that, as Lawrence Lessig writes, negotiating the resulting patent thicket and attendant cost barriers "will entrench the old against the new."[12]

Going forward, we must address the question of whether a fundamental change is necessary in intellectual property law to maintain the ability to innovate. Given the investment in existing patents, is there a way to expand rights for production and use beyond patents in order to facilitate broader participation? One approach is to look more broadly at the way information and goods have been exchanged throughout human history.

Building the Biobazaar?

Among the earliest attempts to explain the phenomenon of open source is Eric Raymond's *The Cathedral and the Bazaar*.[13] The book contrasts centralized control of development projects (the cathedral) with the distributed, voluntary model typified by open-source projects (the bazaar). Explicit in Raymond's telling is the role of individual choice in transactions between leaders of open-source projects and developers, transactions that to Raymond looks like the spontaneous market phenomena observed in a bazaar. The utility and accuracy of the metaphor have been endlessly debated since, but drawing a distinction between organization via centralized control and via spontaneous market phenomena serves as a useful point of departure for thinking about yet another definition of open source applied to biotechnology.

Janet Hope takes the view that the metaphor of the bazaar holds great promise for facilitating innovation in biotechnology. Her book, *Biobazaar: The Open Source Revolution and Biotechnology,* is the first thorough analysis of the social, legal, and economic issues one must consider while envisioning some future version of open-source biology.[14]

Like Richard Jefferson, Hope expresses concern that the patent system is "fully optimized to deliver the maximum amount of socially valuable biotechnological innovation," particularly in the form of drugs and crops.[15] Her larger aim is to describe the circumstances in which for-profit companies might find the open-source model economically useful "without seeking direct compensation in the form of license fees or other pecuniary consideration."[16]

Hope examines many open-source biotechnology efforts presently under way, coming to the conclusion that "open source . . . is not primarily a means of dealing with existing anticommons tragedies. Rather . . . it is a way to preempt such tragedies by establishing a robust commons with respect to basic or fundamental technologies whose value is likely to be enhanced by cumulative innovation."[17]

This version of open-source biology grows organically alongside existing organizations, facilitated by reductions in costs and proliferation of skills and information. These factors are used to justify the adoption of open-source strategies by large organizations—universities, nonprofit laboratories, and companies—for the purpose of facilitating information flow and utilizing surplus capacity and surplus labor to benefit the public good.

Hope observes, "The obvious question for a for-profit company (e.g. a pharmaceutical company) is why give away the crown jewels?" She enumerates several ways that corporations might derive benefit from intellectual property (IP) by "freely revealing," including (1) setting an industry standard, which might lead to growth in sales and may support other products, (2) benefiting from improvements in technology made by others (one motivation for participation in BiOS), which may result in cost savings, and (3) enhancing reputation and brand value.[18]

If even a small number of companies decide to adopt open source as a strategy to reduce costs or outmaneuver competitors, Hope argues, "The actions of those adopters can have a big impact":

> By undermining customers' willingness to pay for access to tools from a proprietary source which they can get at a lower cost from an open one, a small number of open source adopters can shift the balance of competition in a sector away from proprietary technologies. In this sense, open source principles have the power to transform industries. This is why Microsoft is wary of Linux, and it is also why open source has the potential to break the IP logjam in biotechnology.[19]

Hope also notes that an open-source framework need not arise out of whole cloth. While Richard Stallman framed the original open-source license and wrote substantial amounts of code found in today's Linux, his original contribution was turned into a workable whole only through the efforts of others. Stallman's efforts served as a starting point for Linus Torvalds, who supplied the remainder of the initial kernel of Linux.[20] In this case, contributions (of various sizes) from many participants add up to something much larger than the sum of the parts: "Incremental reductions in transaction costs related to intellectual property can have a disproportionate effect on anticommons conditions; there is a kind of tipping point beyond which the payoff for developing an open source substitute for any given toolkit element is high enough for the task to attract focused effort on the part of all who would like to see a fully open collection of tools."[21] Thus as costs continue to fall and participation in developing biological technologies spreads around the globe, we may well see components of a biotech toolkit—an open-source biotech "distribution"—contributed piece by piece from around the world. In fact, this may be the only way to assemble such a toolkit, as people produce and freely reveal work-arounds for proprietary tools, components, and methods that do not by themselves constitute a substantive challenge to existing intellectual property.

Hope is quick to note in other writings that the analogy between software and biology only goes so far, and that open source is not meant to displace existing business models:

> People often conflate the successes and failures of open source software with the possibility of open source biology. . . . It is not necessary to see open source as a stand-alone business model. Instead, it should be viewed as a business strategy that may be used to complement other strategies.
>
> In the end, the proof for the viability of open source biotechnology is not tied to the ultimate success of open source software. Open source software is simply the basis for an analogy—the seed of an idea rather than a rigid formula for success.[22]

That open-source biotechnology was inspired by an analogy with software is a point that deserves careful examination.

Just What Do We Mean by "Source"?

The story of open-source software is fascinating. It has a demonstrably large and growing role in the economy. The idea that software developers could give away the fruits of their labor and still somehow participate in a functioning market is striking and revolutionary. It is also highly distracting.

While it is easy to find a definition of "open source" in the context of software, there are very few examples of firm definitions in the context of biological technologies. If we pursue the strict analogy that DNA is source code, then distributing DNA or its description using an open-source licensing strategy would be the equivalent of building a commons of software. This step requires the right to exclude other uses of the DNA, and U.S. law presently only provides for securing that right via a patent. A list of letters describing a DNA sequences is treated by law neither as code nor as a literary work. As noted earlier, patents are expensive, and obviously protecting a commons consisting of thousands of engineered genes would be prohibitively expensive. Without the ability to propagate low-cost licenses, open-source biology must rely on something else as "source."

The greatest failing of the analogy is that it is solely a suggestive story. As I observed earlier in the chapter, the source code constituting the commons under BiOS licenses is actually intellectual property defined by patents. Because BiOS is a contractual framework built on those patents, all the whys and wherefores are spelled out explicitly, which is both what makes it work and what is likely to inhibit broad use. BiOS is a very concrete and legally defensible contractual framework, but it also incurs substantial costs in maintenance and participation as an extension of existing property-rights schemes. Still, BiOS exists today and is functioning within an extant legal framework. Otherwise, open-source biology or biotechnology as defined by me, by Drew Endy, by Janet Hope, or by anybody else, remains only an analogy—a story—because in no case does it constitute a workable, or testable, economic model.

Thus I will be the first to admit that while "open-source biology" and "open source-biotechnology" are catchy phrases, they have limited content for the moment. As various nonprofits get up and running (e.g., BiOS and the BBF, assuming some sort of license can be agreed upon for the latter), some of the vagaries will be defined, and at least we will have some structure to talk about and test in the real world. When there is a real license for BioBrick parts, and when it is finally tested in court, we will have some sense of how this will all work out.

Why Bother with Open Source?

There are important reasons to try to enable even wider participation in both basic science and in technology development. The open-source software movement is primarily, as Steve Weber argues, "an experiment in political economy." Heavy emphasis should be placed on *economy*, for without an economic impact, whether measured in dollars or productivity or by some other metric, open-source software would constitute simply a collection of code created and traded by software developers in their spare time. And however software developers themselves reap economic benefits from open source—whether via reputation, remuneration, or entertainment—the economy as a whole appears to rely increasingly heavily on that code. Google, with total revenues of just over $16 billion in 2007, is largely dependent upon open-source software.[23]

But it isn't just at large scales that open-source software is influential. Expanding access to code has had substantial economic impacts for individuals and small businesses. Low-cost desktop and laptop computers running Linux are now selling widely in the consumer market, with market share potentially doubling to almost 2 percent in 2008 alone.[24] By lowering the cost of buying and operating a computer, open source has broadened the ability to produce and consume content on the Internet, with the attendant goods of communication and education. Open-source software has also broadened the ability of individuals to write whatever code they desire—for whatever purpose—on hardware running Linux. Open source has become a crucial tool for small-scale innovators.

Can open source similarly be used to facilitate the development of biological technologies? Does the market, as presently constructed, contain mechanisms that serve to promote biological innovation? It may be that, in order to properly deal with innovation in biological technologies, the market requires an overhaul. Alternatively, it may be that the very idea of a market needs to be rethought or that some structure outside the market is required.

Of more immediate concern, is it is possible to employ open-source principles to enable rapid biological innovation, while reducing costs, *within the existing market*? The success or failure of the market to deliver rapid innovation will likely determine our ability to address threats to physical and economic security. The existing market is the environment we inhabit, and this is the environment that provides the context for investment, development, and manufacturing for the foreseeable future.

But that environment is not static, and technology is inexorably altering the ability of innovators to participate in the market. Innovators are emerging at all scales *now,* and we need both to facilitate their work and to construct some means of ensuring that their efforts do not result in unintentional or unrecognized threats. Developing biological analogies to a commons consisting of useful code, and an operating system to run it, holds great promise. The practical implementation, however, does not appear to be forthcoming.

Open Biology

At this point I no longer know what "open-source biology" might be in practice. Biology isn't software, and DNA isn't code—yet. One purpose in writing this book was to open up the conversation and invite as many people as possible to consider where the future is leading us and how we should organize ourselves for maximum benefit.

As in 2000, I remain most interested in maintaining, and enhancing, our ability to innovate. In particular, I feel that safe and secure innovation is likely to be best achieved through distributed research and through distributed biological manufacturing. More general than open source, what we should strive to foster and maintain an "open biology." By "open biology" I simply mean access to the tools and skills necessary to participate in that innovation and distributed economy. The more people who understand the capabilities and limitations of biological technologies at any given point in time, the greater our ability to innovate, to avoid mistakes, to have meaningful public conversations about risk, and to realize the physical and economic benefits of a mature technology. Conversely, as biological technologies become ever more important economically, more people will be motivated to understand how the technologies work, and more people will be motivated to participate in commercial development.

Successful commercialization of biological technologies is highly contingent upon the expected return from investment, which may in turn be contingent upon intellectual property regimes. This is true for large and small participants alike, and it is likely to be true regardless of the financial cost of participating in the market. Even if costs continue to fall, thereby enabling small firms and individuals to invent at will, attempting to market the resulting innovations will be extremely difficult without an ability to secure some sort of exclusive right. Without that exclusive right, innovators are subject to competition by a free rider, someone, as Janet Hope says, "who imitates an invention and thereby gets the full benefit without having made any significant investment of time, effort, skill, or money."[25]

Biological technologies are already so widespread that most products could be easily copied in any country. The global costs of such copying are substantial; approximately 8 percent of world trade consists of counterfeit goods.[26] This is the equivalent of up to $512 billion in lost sales, with about half of that impact on U.S. companies.[27] Thus, Hope notes, free revealing does not presently appear useful in a small biotechnology business: "The fact that an open source license must guarantee licensees' freedom to imitate the licensed technology and distribute it to others means that [the] price tends to be driven down to the marginal cost of reproduction and distribution."[28]

Therefore, if an inventor were to distribute a product via an open-source license—providing access to, say, an engineered protein—the ability to reproduce and distribute the protein might be explicitly communicated in the license. Thus any margin on sales of the protein that might be used to invest in other research and development would be immediately lost to free riders, many of whom may operate at economies of scale that could serve to outcompete the original inventor for sales. The reality of this competitive environment serves to emphasize that the value in many biotechnological innovations is in the design of the object rather than in manufacturing per se. Maintaining the right to exclude competitors from selling artifacts based on that design might be the only way to succeed commercially.

I am not suggesting that work should stop on open-source biology, or on figuring out the relevant licenses, but simply that I do not understand how it aids innovation at the moment.

A great deal of the innovation that our society and economy needs will not come from academia or existing corporations but from people noodling around in their garages or in start-up companies yet to be founded. These are the important users of BioBrick parts over the long term; these are the people who need access to a licensing structure that does not require a patent in order to commercialize further innovation; these are the people who want the ability to build biological systems without needing a large government grant. As costs fall, people will follow their interests and participate however they are able.

Regardless of how a license is structured and whether a project can be considered officially "open source," a substantial challenge in keeping biology open is maintaining the ability to choose what to work on and how much effort to supply. Moreover, from the perspective of project management, it is important to recognize that experience and expertise are never distributed uniformly among the participants in any project. Skills and commitment often differ wildly, whether in software development, basic science, policy development, or engineering. It is also true that we can never be entirely sure who will contribute a crucial piece of the puzzle. We must be sufficiently inclusive to catch all the bright ideas; we are all customers and participants in open biology. The economic structure of that participation requires more attention.

Both Drew Endy (of the BioBricks Foundation) and Richard Jefferson (of CAMBIA) presently have as primary customers not hobbyists, tinkerers, or engineers, or even a large number of corporations, but rather students, foundations, and governments. The BBF and CAMBIA are themselves nonprofit organizations funded thus far primarily through grants and donations, not fees, with stated goals of expanding access to technology rather than producing and accumulating capital. The marketplace in which these organizations compete for funding values invention, innovation, and the promise of changing the world. At present, the BBF and CAMBIA do not derive the majority of their funding from actually selling parts or open-source licenses on the open market, and thus they do not rely on sales to fund their work—*nor should they*. If the BBF and BiOS can make their respective models work for themselves and their existing customers, then more power to them. But the rest of our economy operates on exchanges of money for goods and services.

It could be argued that the present structure of innovation in biological technologies works just fine: you have an idea, you secure initial funding, pay for a patent, and then raise more money. This style of innovation relies very heavily on the availability of funding by venture capital, which has certainly produced remarkable results over the years in Silicon Valley. However, not all venture capitalists (VCs) have the golden touch, and only a fraction of venture-funded companies become truly successful. Those firms go on to large sales volumes, public stock offerings, and perhaps acquisition by larger firms, generating returns that cover losses on the other VC investments. Based on evidence from Silicon Valley, this is evidently a reasonable way to make money for VCs. But this way of doing business constitutes a terrible strategy by which to build a technology base. Patents owned by companies that go bust, or even by those that survive but go nowhere financially, do not serve the public good and may wind up being acquired and used to suppress innovation by third parties. The way we finance innovation in biological technologies clearly needs some adjustment if we are to maximize the utility of those innovations. Alternatively, if the cost of innovation continues to fall and if more entrepreneurs begin practicing garage biology (see Chapter 12), then the VC model may find itself outcompeted in the market it helped create.

Looking ahead, we must conceive of a financial and legal system that (1) enables strong protection of individual parts when those parts are economically valuable if used by themselves and (2) simultaneously facilitates broad licensing of the parts when economic value comes from combining many parts into a larger whole. These two different requirements support very different business models, which only serves to complicate the situation. As described in the previous chapter, when it comes to patent reform, large biotech companies concerned about patents on individual genes and molecules are at odds with technology companies that require access to large numbers of parts.

Eventually we must create a system that facilitates granting an exclusive right for *complex* biological circuits, a system that individual "developers" can afford to utilize. This requirement is as much a part of open biology as is maintaining access to technology; clearly the ability to specify the use of new designs has had a profound effect on software innovation. There is presently no obvious path to creating such a system for biological innovations other than relying on patents, but a lesson from another segment of the economy might provide inspiration.

Listening to the Source

An excursion into an entirely different realm of innovation and property can help us to reimagine these concepts in biology. David Byrne notes,

> In the past, music was something you heard and experienced—it was as much a social event as a purely musical one. . . . It was communal and often utilitarian. You couldn't take it home, copy it, sell it as a commodity (except as sheet music, but that's not music), or even hear it again. Music was an experience, intimately married to your life. You could pay to hear music, but after you did, it was over, gone—a memory.
>
> Technology changed all that in the 20th century. Music—or its recorded artifact, at least—became a product, a thing that could be bought, sold, traded, and replayed endlessly in any context. This upended the economics of music.[29]

For most of human history, there was no way to preserve a copy of a performance on physical media. You could remember what you had heard, or you might read sheet music and hear the notes in your head, or you might play an instrument or sing; following a loose analogy, we can say that the "source code" of music was the performance, with sheet music or memorization used as an aid to remember which note went where, and when. Sheet music embodied the only physical representation of music beyond experiencing it in person.

From this perspective, sheet music is merely a storage medium and, like other uses of paper, is being displaced by silicon, plastic, and glass—the stuff of digital storage. Similarly, the invention of the piano roll and, later, the gramophone created the ability to store the "source code" of music on physical media in the form of a "program" that could be run via mechanized instruments. Radio later served as a distribution channel for both recorded and live performance.

The appearance of recordable media gave rise to the opportunity to capture economic value from the listener's desire to hear sounds displaced in time and distance from where those sounds were originally produced. The process of producing and distributing recordings originally required substantial infrastructure and capital, and a new industry grew up to support the activity.

During the twentieth century, the economics of the music industry was dominated by the costs of recording, distribution, and marketing, all of which benefited from scale. As presented in Chapter 10, industries in which costs are predominantly in production tend to be dominated by large firms that can afford to support such large-scale activities.

This model is clearly no longer working in the music industry. Sales are declining precipitously, and a rapidly growing fraction of those sales (now almost half) are digital.[30] At the moment, contracts between labels and artists tend to treat downloads as albums, with record companies receiving the same fraction of the purchase price from the electronic copy as from the physical one, despite the reduced production costs.[31] Clearly, that can't last.

With the emergence of new technology, the relevant scale in the music industry is changing. The ability of middlemen to capture substantial value in recording sound on physical artifacts appears to have been transitory. As industry sales have fallen, the centralized production and distribution of physical media is losing its appeal as a competitive destination for capital. Technology enabled the existence and growth of that segment of the economy, and technology is now reducing it. Technology has once again changed the value proposition. The creation of economic value has been redistributed to a smaller scale.

Anyone can now try his or her hand at making and marketing music. The costs of recording, reproducing, and distributing both physical media and electronic copies have dropped to near zero.[32] Many artists are already taking on considerably more of the recording and production of their music, forcing record labels to find a new role. Where once the corporate entity, in effect, owned the music and marketed the artist as a brand, that is now just one end of a range of different kinds of deals, with a do-it-yourself model at the other end. In between are arrangements that involve licensing, publishing, or profit sharing or that may be restricted to manufacturing and distribution. This variety of economic arrangements is beginning to look a lot more like the range present in the rest of the economy. In other words, as the cost of participating in the market and reaching consumers has fallen, the monolithic model of dominant record labels has fractured into a more-complex landscape with many different niches.

With the increase in electronic transfer of music via the Internet, the music industry lobbied heavily for the passage in the United States of the Digital Millennium Copyright Act (DMCA). The DMCA increased penalties for infringing on copyright by electronic transfer of files and expressly made illegal the circumvention of security measures on digital copies of copyrighted materials.[33] It isn't yet clear whether the DMCA will truly aid the economic fortunes of the recording industry—piracy is still rampant, and artists appear to be increasingly seeking their own direct connections with fans—but the DMCA did substantially alter the legal, and thus economic, impact of information traffic.

The relevance of these changes to the future of biological technologies is twofold. First, intellectual property laws are not set in stone. Laws can be changed to reflect advances in what is considered art and design or changed in response to the emergence of new technologies that either threaten old value chains or that open up the possibility of new ones. While copyright presently covers content and patents presently cover function, this situation need not be permanent.

Second, the very notion of what is considered "product" in the music industry is changing, with the value now being placed more on the experience of the listener. Similarly, as costs fall, reconstituting DNA sequences from electronic information, and using those sequences to control systems and produce objects, will become a smaller fraction of the overall costs. DNA sequencing and synthesis have already become commodities. The design process and the description of the resulting parts and circuits will soon be the most valuable portion of building new biological artifacts. The "source" in this case, the thing that people will care about and that will have economic value, will become electronic information that specifies biological function.

Technological changes have fundamentally upended the structure of the music industry, turning the maintenance of large-scale assets and organization that formerly served as an advantage into a liability. This is, despite the wide role of music in human civilizations and economies, a trivial change for society as a whole. Similar changes in other technologies, in contrast, are bound to have a more-significant effect. We are living through an era when technology is putting more power into the hands of individuals than has ever occurred in the history of human civilization.

The Revolution Is behind Us

The global proliferation of the capabilities to communicate and to innovate will have profound impacts on the structure of production and therefore on the structure of the economy. The means presently used to grant an exclusive right to the resulting works of art and science may be outmoded already, with the realization simply yet to set in. It may be time for a change in the way exclusive rights are granted.

Neither patents nor copyrights are specified in the U.S. Constitution. Both patents and copyrights are derived from European practices many centuries old. Consequently, neither the legal nor the economic frameworks governing intellectual property constructed in the sixteenth through twentieth centuries appear well suited to a world in which bits and atoms are fungible. Increasingly, the methods used to arrange atoms and set them in motion will be more diversified and distributed, with the instruction set stored and transmitted as bits through a wide variety of means. Patents are obviously crucial to me as an entrepreneur right now: in ten years, maybe not so much. The present structure of intellectual property law around the world could well fall behind technology even further than it already has.

It is possible that the strategy presently being pursued by the BioBricks Foundation of putting parts into the public domain will slowly come to dominate the innovation landscape. Simply opting out of the existing intellectual property framework might constitute a socioeconomic hack analogous to copyleft and might serve as a "tipping point," after which owning methods and objects becomes less important than access to information.

Technology is advancing to the point where design information for biological artifacts may become more relevant for physical and economic security than the objects those designs describe. The economic value of centralized production rests on the competitive advantage of a third party to produce objects for the consumer. If centralized production becomes less important through the fungibility of matter and information, then value becomes concentrated in information.

The bigger issue, then, is that protecting the value of information in general is difficult when the infrastructure already exists to reproduce it at will. On the time scale of just a few years, this phenomenon will strongly affect the economy of biological technology, just as the descendants of Napster are breaking open peer-to-peer data-sharing today. In a relatively short period of time, that is, relative to the lifetimes of corporations such as IBM, DuPont, and Novartis, instruments capable of sequencing, resynthezising, and thus reverse-engineering any DNA-based technology (that is, life as we know it) will be present on lab benches around the world and quite probably in many garages and kitchens. In this world, how will any organization, commercial or otherwise, survive by relying on proprietary information?

Extending the argument a bit further, one must ask what will happen to commerce when information is sufficient to reproduce biological systems. What value will a patent have if no commerce exists around the objects or processes claimed as inventions? What utility will patents on biological technologies have? And how will those patents contribute to the public good? Despite the appearance of bestowing ownership, patents provide no ownership of things or processes but rather confer the temporary right to exclude anyone else from benefiting economically from their sale.

When it eventually becomes possible to synthesize DNA at will for minimal cost and to run that "program" in an organism of one's choice, there may very well be no market for the object defined by the program. In that world the value of DNA truly becomes its informational content. The only money that would exchange hands in a transaction would be to pay for the DNA synthesizer and for reagents and raw materials to run the synthesizer and other instruments.

As is quite clear from earlier chapters in this book, the technology required to implement this infrastructure does not yet exist. It should also be clear that, at least for simple implementations, it isn't so far away. The difference between that world and the present is one of degree rather than of fundamental shifts in technological capability. We do not need to find a way to force open the doors of innovation; they have been slowly opening, though occasionally stalling, for more than a century.

If the analogies presented in Chapters 1 through 6 between automobiles, aviation, computers, and biology teach us anything, it is that once the process of developing an engineering discipline in a field of human endeavor gets started, and particularly once it begins demonstrating economic value, it cannot be stopped. Progress may be slowed by technological difficulties, and the Technology Line presented in Chapter 5 points toward the kinds of investments that should ease the transition to real biological engineering. Substantial work on the required foundations is already underway (see Chapters 6 through 8). Synthetic biology and related technologies appear to be crucial elements of dealing with natural and artificial threats, as illustrated in Chapter 9. Producing the innovation required to meet those threats requires enabling participation of individuals and small firms, and though prohibition in the name of improving security may temporarily limit access to hardware and may temporarily restrict the practice of certain skills, the strategy appears historically futile and will in fact reduce security in short order (see Chapters 9 and 10). Both physical and economic security are best guaranteed through diverse and open innovation, which will further contribute to economic growth (see Chapters 10 and 11). Progress may be slowed by squatting on patents to the detriment of the public good (see Chapters 12 and 13), but regardless of the terms under which intellectual property is shared, or stolen, we have already witnessed a fundamental change in the structure of innovation and production.

We cannot go backward. The world has already changed irreversibly, and while specific economic and social impacts are already becoming apparent, a wider appreciation of this change has been slow to dawn. The question we must face is how to go forward. The revolution is behind us.

CHAPTER FOURTEEN

What Makes a Revolution?

SOMETIMES REVOLUTIONS APPEAR IN HINDSIGHT, the outcome of upheaval you didn't see coming, the result of change nobody managed to trumpet and claim credit for. There is currently a revolution well under way, progressing quietly amid the cataloging of parts and the sequencing of bases. We are only now beginning to understand the power at our fingertips.

Recall that products derived from modified genomes are already the equivalent of about 2 percent of U.S. gross domestic product, with an absolute monetary value that is growing at 15–20 percent per year. In the period 2006–2007, the monetary value of industrial and agricultural applications of biotech surpassed that of health-care applications. This in itself might constitute a revolution, demonstrating a growing demand for products made using biological technologies, products that must have demonstrably useful and reliable behaviors in order to compete in the marketplace.

The connection of biological technologies with markets is crucial, because without demand there will be no investment that enables conversion of inventions into innovations. In a world in which products are made via biological manufacturing, the diffusion of innovation will almost certainly be entangled with novel economic and political structures.

What kind of political and economic technology will influence, or be influenced by, ongoing upheaval in the development and application of biological technologies? This question points the way to thinking very broadly about revolutions. Examining revolutions within specific historical contexts helps elucidate both the various contributing factors and whether the participants knew what was happening at the time. Among the best-studied revolutions are the military sort, which serves as an interesting place to start thinking about current events.

The Opportunity and the Threat in the Military Use of Biological Technologies

Any treatment of the future of biological technologies would be incomplete without examining potential military applications. It would be naive to expect that militaries around the world would *not* be interested in employing new biological technologies, because military organizations always exploit the new to gain advantage over adversaries. It is also the case that for biological technologies in particular, it behooves all observers to closely examine just how military and security organizations invest in deploying biology, and to what ends.

The first military application that most people imagine for biology is the use of pathogens as weapons. Weaponization of biological technologies and their use to develop new pathogens simply to study their properties are, for me, by far the most frightening prospects of the coming decade. We should distinguish, however, weaponization or biodefense applications from the use of biological technologies in support and production.

Certain military uses of biological technologies could serve several beneficial ends. The general present concern about bioterrorism has led to an important investment in new detection instrumentation, which could eventually lead to civilian applications for diagnostics in health care and to engineering tools for biological systems.

Stepping even further away from either weapons or countermeasures, military adoption of biological technologies in logistical and procurement transformations could save substantial sums of money and dramatically reduce the environmental impact of military operations. The biological production of aviation fuel would simply bring military uses of biological technologies in line with the rest of the economy. Within the U.S. Department of Defense (DoD), the Air Force is the largest consumer of fuel, and every dollar increase in the price of a barrel of oil raises annual costs by $60 million.[1] Military aircraft in all service branches account for 73 percent of DoD fuel use, and each dollar increase in the price of oil raises overall costs $130 million.[2] Price fluctuations are, obviously, extremely expensive for the DoD and are hard to incorporate into planning. Thus the prospect of a jet fuel replacement produced biologically at the equivalent of $50, as described in previous chapters, would represent both a multibillion dollar savings for military operations and a dramatic improvement in the stability of projected costs. Similarly, if it were a country in its own right, the Department of Defense would rank thirty-sixth in total fuel use among nations worldwide, which makes the DoD a large emitter of greenhouse gases; a carbon-neutral fuel supply would substantially reduce the consequent environmental load.[3]

A significant military investment in biological technologies would, in effect, bring an enormous financial lever to bear on developing infrastructure and economies of scale. The history of computers and aviation make clear the important historical role of military research-funding and procurement to guarantee investment in production, thereby giving whole new industries a firm footing.

Pursuing the development of biological technologies to reduce military maintenance and operational costs is the result of a choice. It is a choice available to many military organizations around the globe, but it is not, at least at this point in time, a required course of action that must be taken in order to guarantee the survival of any country or its soldiers. The presence of a choice and its availability to many organizations around the globe are factors historians use to help understand the course of conflicts.

Williamson Murray and MacGreggor Knox, both veterans of combat in Vietnam, distinguish great "military revolutions," in which an "uncontrollable rush of events sweeps nations and military organizations before it," from "revolutions in military affairs" (RMAs), which "appear susceptible to human direction."[4] Elaborating on this difference, Mark Grimsley writes, "Individuals or groups do not control military revolutions: they merely seek to survive them."[5] Military revolutions are therefore phenomena that transcend militaries to encompass economies and societies. Table 14.1 gives examples of five military revolutions from the last four centuries.[6]

Murray and Knox argue,

> These five upheavals are best understood through a geological metaphor: they were earthquakes. They brought systemic changes in politics and society. They were uncontrollable, unpredictable, and unforeseeable.
>   Military revolutions recast society and the state as well as military organizations. They alter the capacity of states to create and project military power. And their effects are additive. States that have missed the early military revolutions cannot easily leap-frog to success in war by adopting the trappings of Western technology.[7]

Appreciating the difference between military revolutions and revolutions in military affairs is aided by a more-detailed elaboration of how the five military earthquakes elucidated above can be broken down into contributing

TABLE 14.1   Five military revolutions from the last four centuries

---

Military Revolution I: The creation in the seventeenth century of the modern nation-state, which rested on the large-scale organization of disciplined military power.

Military Revolution II: The French Revolution of the late eighteenth century and after, which merged mass politics and warfare.

Military Revolution III: The Industrial Revolution of the late eighteenth century and after, which made it possible to arm, clothe, feed, pay, and move swiftly to battle the resulting masses.

Military Revolution IV: The First and Second World Wars, which combined the legacies of the French and Industrial Revolutions and set the pattern for twentieth-century war.

Military Revolution V: The advent of nuclear weapons, which contrary to all precedent kept the Cold War cold in the decisive European and northeast Asian theaters.

---

factors. Table 14.2 is adapted from a list presented by Murray and Knox; in all cases military revolutions and the associated RMAs consist of intertwined technical, economic, and social changes. Thus, technological developments do not by themselves constitute a revolution, military or otherwise.

Introducing the jargon of military revolutions and RMAs helps to shine an explicit light on the various kinds of direct military use of biological technologies. Pathogens have been used as weapons for most of recorded history, and modern states have previously developed biological weapons. Such activity at the state level appears to be halted, for the moment.

To the extent that biological technologies could be used to pose a military threat, they must be placed in the larger context of social and economic changes. Assuming that sanity prevails and that states and their militaries refrain from developing biological weapons, biology merely becomes a piece in a larger puzzle. As explored throughout this book, the diffusion of biology as a technology throughout the economy depends on the interaction of social, market, and regulatory factors. Knox and Murray note that "changes in society and politics are the most revolutionary forces of all."[8] The emergence of new technologies and new kinds of social organization suggest that the current revolution be broadly defined.

The United States is among many countries presently engaged in a "war," not against any state in particular, but against empowered individuals and groups. The vernacular of the day makes this a fight against "terrorism," defined as a strategy to create economic disruption and fear for

TABLE 14.2   Military revolutions and associated revolutions in military affairs

---

Military Revolution I: 17th C. Creation of modern state and modern military institutions
Associated RMAs:
   1. Dutch, Swedish, French tactical reforms of the battlefield organization of troops;
   2. Naval revolution for projection of force at sea;
   3. Britain's financial revolution.

Military Revolutions II and III: French and Industrial Revolutions
Associated RMAs:
   1. National political and economic mobilization, origination of nationalism and national identity; Napoleonic warfare designed to obliterate enemies, broad education and training of officer corps;
   2. Financial and economic power based on industrialization;
   3. Technology revolution in land warfare and transport (telegraph, railroads, steamships);
   4. Big-gun battleship and battlefleet.

Military Revolution IV: WWI and WWII. Combined preceding revolutions
Associated RMA:
   1. Combined-arms tactics and operations, Blitzkrieg operations, strategic bombing, carrier and submarine warfare, radar, signals and intelligence.

Military Revolution V: Nuclear weapons and ballistic missiles
Associated RMA:
   1. Precision reconnaissance and strike, stealth, computer communications and control, massively increased lethality of "conventional" munitions.

---

the sake of publicity or the purpose of forcing policy changes. That definition of terrorism is too narrowly circumscribed both historically and operationally.

Jamais Cascio, a scenario planner, futurist, and both a student and a proponent of open information economies, considers present challenges to be examples of a longer-term emergent trend: "Conventional military forces appear to be unable to defeat a networked insurgency, which combines the information age's distributed communication and rapid learning with the traditional guerrilla's invisibility (by being indistinguishable from the populace) and low support needs. . . . Insurgencies have always been hard to defeat with conventional forces, but the 'open source warfare' model, where tactics can be learned, tested and communicated both formally and informally across a distributed network of guerillas, poses an effectively impossible challenge for conventional militaries."[9]

Whether the challenge of "open source warfare" is "impossible" to meet or not remains an unfinished experiment. Yet regardless of the cause of the present conflict, and regardless of the political consequences of any particular tactic or strategy employed in the course of fighting, the scale and cost of the conflict appear to mark it as a new military revolution. So-called illegal combatants, who are not soldiers in the service of any recognized state but rather profess allegiance to particular ideals and ideologies, make use of modern communication and transportation and demonstrate a willingness to use civilians and infrastructure as both weapon and target. They employ tactics and technology that, at least in combination, constitute new threats for national military organizations. Common materials and consumer devices, combined into "improvised explosive devices" (commonly referred to as IEDs), have been used to cause substantial casualties to ground forces in Iraq and Afghanistan and have necessitated dramatic shifts in tactics and an extensive ongoing effort to develop countermeasures by the U.S. Department of Defense.[10]

While combat continues in Afghanistan and Iraq, many thousands of civilians worldwide have been killed over the last decade in attacks that employed readily available materials. Instructions to build bombs using common materials are widely available on the Internet, and instruction in person appears widespread around the world in training camps. The availability of relevant material and information are a feature of our economic and social system that would be difficult, if not impossible, to alter, even if the political will existed to take such action. Although bases and training camps may be shut down and individuals may be placed under surveillance or detained, the larger market for, and flow of, information and material around the globe is here to stay. Biology is simply another factor in play. Fortunately, biological technologies have thus far not been deployed in recent conflicts beyond the distribution of anthrax spores via mail within the United States in late 2001. The latter now appears to have been an isolated act of domestic terrorism.[11]

It is not obvious that any biological technology, and gene synthesis in particular, provides capabilities that usefully extend the reach of empowered individuals. By itself, gene synthesis constitutes a powerful technology that is presently beyond the reach of all but a few skilled in the art. However, when combined with design capabilities, electronic communication of sequence information, and worldwide express delivery of objects of all sizes, gene synthesis is becoming an accessible and widely useful tool.

Looking forward, as skills and knowledge become more distributed around the world, gene synthesis will eventually become less reliant upon centralized production. This will come to pass even if the use of DNA synthesizers and access to gene-synthesis technology were proscribed, as described in previous chapters, because individuals with sufficient knowledge could fabricate the necessary instrumentation using readily available parts and information. Thus the broader present revolution appears to be defined by a combination of (1) widely distributed knowledge, (2) rapid communication, and (3) access to parts and fabrication facilities that can be used to assemble weapons or other technology on a custom basis. For example, while it does not appear that IEDs by themselves threaten the viability of any state or military organization, the larger communications and industrial capacity that enables IEDs, and associated strategy and tactics, are developments that no nation can afford to ignore.

The seeds of the present technological and economic upheaval actually germinated gradually over the last three decades, relatively far removed intellectually from the biology of the day. As introduced in the previous chapters, the emerging concepts of open source, open innovation, and peer production fundamentally alter the role of individuals in generating and propagating innovation. I have gathered together many technologies and activities that I see as contributing to this revolution in table 14.3.

Some aspects of economic, technological, and political change appear controllable. We experience, or suffer, other aspects simply because they are too large or complex to manipulate. Not all revolutions are voluntary, nor must they be sudden. Sometimes revolutions sneak up on us; often they are evident only in retrospect.

The Revolution of Open and Distributed Innovation

If the definition of a military revolution has in the past been that missing one means you cannot "easily leap-frog to success" by buying in at a later date, then we may find that the factors listed in table 14.3 define a new form of revolution. In this kind of revolution, material and information are fungible, infrastructure is easy to build, and skills are easy to obtain. For example, the U.S. military has become heavily reliant for surveillance on unmanned aerial vehicles in Iraq and Afghanistan. However, the mili-

TABLE 14.3   Revolutions in Military Affairs contributing to a twenty-first-century military revolution

Military Revolution VI: Empowered non-state groups and individuals, the fungibility of matter and information, distributed and open innovation

Associated RMAs:
1. Worldwide electronic communication, surveillance, and finance;
2. Agreements between states do not limit, and perhaps incite, individuals or groups willing to take action;
3. Individuals willing to take own lives, use of civilian assets in attacks, large-scale attacks on civilian assets as tactical terror targets rather than strategic wartime targets;
4. Goal of attacks may be actual physical and/or economic damage, or simply to cause fear;
5. Ready access to Chemical & Biological (C&B) materials, ready access to infrastructure and skills to manipulate C&B materials;
6. Ready access to mechanical and electronic components, ready access to information and computer aided machine tools, all in the context of a burgeoning "do-it-yourself" hardware hacker culture;
7. Design information and physical objects are interchangeable for the purposes of commerce and war.

tary's foes on the ground have learned to evade such monitoring during the planting of IEDs, and the means of surveillance itself has become the object of widespread innovation.[12] The U.S. military could rapidly find itself subject to the same sort of aerial scrutiny it has introduced into the battlefield in the hopes of gaining tactical advantage.

Remote control airplanes are a subject of great interest to hardware hackers. There is a worldwide amateur effort, coordinated online, to share information about building better control systems, power systems, autopilots, GPS route–following capabilities, and, of course, aerial reconnaissance systems.[13] Anyone who wants to learn from the effort can do so. The community is aware of potential nefarious exploitation of their explicit inclusivity. The general feeling among participants appears to be that sharing information is better than hoarding it and that they are helping to keep government authorities apprised of what is technically feasible for educated and motivated amateurs.[14]

The larger hardware-hacking community is actively developing much more general fabrication capabilities, with three-dimensional printers, machine tools, and laser cutters all employed in the service of building cool stuff. Results, methods, and designs are often published in the public domain. The MIT Fab Lab project has assembled a list of machine tools that can be purchased for a relatively small sum and that, used together, enable building a wide variety of functional objects:

> [A Fab Lab] is a group of off-the-shelf, industrial-grade fabrication and electronics tools, wrapped in open source software and programs. . . . Currently the labs include a laser cutter that makes 2D and 3D structures, a sign cutter that plots in copper to make antennas and flex circuits, a high-resolution milling machine that makes circuit boards and precision parts, and a suite of electronic components and programming tools for low-cost, high-speed microcontrollers. . . . A full Fab Lab currently costs about $25,000 in equipment and materials without MIT's involvement.[15]

Partner Fab Labs have thus far been set up in Lyngen Alps, Norway; Cartago, Costa Rica; Pabal, India; Boston, Massachusetts; Jalalabad, Afghanistan; and several sites in Africa. With training, anyone is free to use a Fab Lab to build what he or she wants, or just to learn. A similar project, Fab@Home, aims to foster the development and use of three-dimensional printers— "fabbers"—capable of fabricating everything from confectionary flowers to functioning electronics.[16]

Plans, parts lists, and software to run fabbers are available online, with kits and assembled printers available from commercial vendors. High-end fabbers originally intended for commercial use have fallen so far in price that they are expected to become consumer items shortly, enabling the printing of metal and plastic objects at home; writes the *Economist,* "The real fun will come as ordinary folk at home feel free to let their creativity run wild. If you can imagine it, you can make it."[17] One goal of participants in both the Fab Lab and Fab@Home is the eventual creation of tools that can be used to reproduce themselves, limited only by access to plans and raw materials. The RepRap is advertised as an early version of a "self-reproducing printer," for which the source code is distributed via the Internet under the GPL (see Chapter 12).[18] It is worth mentioning here that the Fab Lab in particular appears to constitute a set of tools entirely sufficient to build from raw materials all the parts required to assemble a conventional DNA synthesizer.

The lesson is very general: there are no elements listed in table 14.3 that appear restricted to the use of militaries or illegal combatants. That is, the specific elements of modern communication and fabrication that are presently forcing the largest military organization on the planet to rapidly innovate, in order to adjust to new combat realities, are part of a much larger economic revolution. The present economic revolution might be defined as one that affects both the structure of innovation and the means of production. This alters the capacity of states, organizations, and individuals to participate in the modern world and to provide for the needs and aspirations of their allegiants.

The Larger (Scale) Context

Information flow is a foregone conclusion. There is little that can be done today to curb that flow; this situation arrived before we—the general public—were aware of it, and our dependence on such technology has made it indispensable to the physical and economic security of all states and organizations. Cell phones are present even in the poorest corners of the poorest countries on the planet, and Internet access is following rapidly. The consequent improvement in communication has facilitated education, transfer of capital, and distribution of price information that allows remote farmers to know what to plant to get the best return at market.

We are also witnessing a general and distributed improvement in the ability to control matter. Fab Labs and open-source three-dimensional printers are at the cutting edge. But as even conventional manufacturing plants move around the globe, they leave skills and infrastructure in their wake. Countries that used to be destinations for factories solely because of cheap labor now contribute their own new products to the global market. As V. Vaitheeswaran notes in the *Economist,* this is an iterative process: "Chinese and Indian companies can practise on their domestic customers while they improve quality to the point where they can begin to export. South Korean firms have already gone through much the same process with consumer-electronics and cars—and in the process have frightened many of their Japanese rivals."[19] Geographically distributed increases in the ability to innovate are occurring while innovation itself is speeding up. A study by Procter & Gamble found that the life cycle of consumer products fell by half between 1992 and 2002.[20] Independent of whether a state or organization is large or small, keeping up in this world requires access to tools and skills that enable rapid innovation. This means not just having access to communication and fabrication as a tactical necessity but also improving those capabilities as a strategic goal. Therefore, any policy choices implemented with the aim of increasing physical and economic security must acknowledge the reality of pervasive communication and the fungibility of bits and atoms. The new economy of bits and atoms is explicitly and irreversibly transnational.

Only time will tell what effect communication and distributed fabrication will have on the scale of innovation and production using biological technologies. Even high-end laboratory instrumentation for DNA sequencing and synthesis already exists in a complex marketplace with multiple competitors. There do not appear to be any advantages for large-scale design and assembly operations; specialized services such as gene synthesis have rapidly become commoditized through the emergence of many competitors. Even if the commodity nature of gene synthesis drives consolidation, there is no evidence that capital requirements will be a significant barrier to entry for competitors at any time. This is particularly true if costs continue to fall exponentially and automation becomes more useful and cost effective (see Chapter 6). Similarly, as production pathways for drugs, chemicals, and fuels are embedded within organisms or in vitro systems, the skill and cost barriers to using and reprogramming biological systems will fall (see Chapters 7, 8, and 11).

In the context of low barriers to entry and low production costs, one must ask what advantage remains for Goliath. The answer may be in branding, quality assurance, and accumulating intellectual property.

The recipe for making acetylsalicylic acid‚Äîaspirin‚Äîhas been in the public domain for many decades and can easily be followed in the kitchen or garage to produce the molecule for medicinal use. Yet despite widely available recipes and reagents, everyone buys prefabricated aspirin from the local pharmacy, grocery store, or gas station minimarket. The obvious reasons for this are a combination of cost, effort, knowledge, and some guarantee of safety. As a U.S. Small Business Administration report notes, "customers reward established firms that can reliably provide products and services with known attributes." This reliability does come at a cost, namely, that it "reduces adaptability, because it is achieved by reducing variation in the organization's activities that otherwise would have provided opportunities to innovate, in order to satisfy expectations of existing customers."[21]

In the pharmaceutical market, existing companies clearly have an advantage when it comes to the power of brands. They also tend to have plenty of cash and contribute substantial amounts of that cash to new research and development. Unlike the majority of mature industries (see Chapter 11), innovation in the drug industry‚Äîwhether in producing small molecules or biologics‚Äîis presently dominated by large companies.[22] Despite spending more than $50 billion a year on research and development, the *Economist* notes that "pharmaceutical giants continue to get their hands on new science by buying small innovative firms, particularly in biotech."[23] But a small biotech company is not the same thing as a small software or hardware company.

When innovating by acquisition, Big Pharma tends to shop for companies with promising products already well into the formal trials to test safety and efficacy. To reach this stage of drug development, these "baby biotechs" have usually raised and spent at least tens—if not hundreds—of millions of dollars, and companies with products on sale in the research and diagnostic markets may already have substantial revenues. Biotech firms that make it all the way to market spend about as much to develop a new product as do major pharmaceutical companies, between $800 million and $1 billion.[24] This is not exactly the atmosphere breathed by the average entrepreneur. And despite the dollars spent by pharmaceutical companies and the baby biotechs purchased, the productivity of pharmaceutical research dollars is falling (see Chapter 11).[25]

Gary Pisano, a professor at Harvard Business School, asserts that the problem is structural. Raising the question, Can science be a business? he argues,

> The "anatomy" of the biotech sector—much of it borrowed from models that worked quite well in software, computers, semiconductors, and similar industries—is fundamentally flawed and therefore cannot serve the needs of both basic science and business.
>
> By "anatomy," I mean the sector's direct participants (start-ups, established companies, not-for-profit laboratories, universities, investors, customers); the institutional arrangements that connect these players (markets for capital, intellectual property, and products); and the rules that govern and influence how these institutional arrangements work (regulations, corporate governance, intellectual property rights).

Pisano contends that, in the face of the financial and regulatory burdens of developing applications for manipulating complex systems, successful innovation requires a different approach. His diagnosis is long and detailed, but Pisano explains what he sees as the poor performance of biotechnology through the "profound and persistent uncertainty, rooted in the limited knowledge of human biological systems and processes, [which] makes drug R&D highly risky."[26] Combined with the complexity of human biology, the propensity of universities and businesses to patent everything in sight (see Chapter 12) is creating a thicket of intellectual property that is hard to consolidate for the purposes of developing drugs. Among Pisano's recommendations are implementing greater vertical integration of research, development, and manufacturing and reducing the number of independent biotech firms. That is, Pisano is arguing that innovation is best served in drug development by even larger-scale companies than already exist.

Regulation and market complexity are issues that are certainly worse in biotechnology than in other areas of technology. But the argument for seeking greater innovation by turning to larger-scale organizations seems to run counter to the history of the last several centuries (see Chapters 5 and 10). Undoubtedly the causes of the declining value of the research dollar are both many and complex, but William Baumol's division of labor in innovation provides an interesting perspective on the role of entrepreneurs seeking to develop new biological technologies (see Chapter 10).

Within the context of human health-care and drug development, the David and Goliath model may suggest a change in operating procedure for pharmaceutical companies. Big Pharma might find its R&D dollars better spent to reach further down into the innovation pipeline, spending less on each of a larger number of innovations. If, in the end, scale is the only way to deal with issues of developing effective and safe therapies, then one must also remember that biotech drugs now constitute less than half the total revenues generated by biological technologies (see Chapter 11). It is not necessarily clear that biotech Goliaths have advantages over the Davids in areas outside of health care. And it is in considering the rest of the bioeconomy that distributed innovation, pervasive communication, and the fungibility of bits and atoms becomes most important.

"The Shock of the Old"

Novel artifacts and technologies do not arrive cut from whole cloth: they cannot. We must use existing tools and concepts in the process of invention and innovation. We build the new from the old. Similarly, as historian David Edgerton writes, we must maintain the old: "The innovation-centric view misleads us as to the nature of scientists and engineers. It presents them, as they present themselves, as creators, designers, researchers. Yet the majority have always been mainly concerned with the operation and maintenance of things and processes; with the uses of things, not their invention or development."[27] The view that understanding the development of technology requires looking backward as much as forward and looking at use as much as at innovation is a key argument made by Edgerton, of Imperial College London, in his book *The Shock of the Old*. Looking backward reminds us that, in all cases, substantial numbers of people are involved in keeping the physical and electronic infrastructure of our socioeconomic system from crashing.

Even those scientists and engineers who do participate in invention and development will recognize that much of their time is spent in maintaining the equipment they build, buy, and use. It is usually only by looking back over time that the costs of maintenance become apparent. While these costs are sometimes considered in the purchase of specific consumer goods, commercial equipment, and manufacturing systems, it turns out that maintenance costs are rarely included in estimates of GDP.

Canada is among the few countries that keep specific statistics on maintenance, which amounted to a healthy 6 percent of Canadian GDP in the years 1961–1993.[28] Edgerton cites many isolated examples of such data, where maintenance costs range from 50 percent to more than 100 percent of the original investment in roads, buildings, and manufacturing and mining equipment. In the 1980s in the United States, building renovation and repair amounted to 5 percent of GDP, one-and-a-half times that spent on new construction.[29]

New industrial processes are all constructed by adapting and modifying old ones, and we step forward gradually as we understand new applications for extant technology. Therefore, in order to construct better narratives about technological change, we must consider the present as the midpoint rather than the beginning.

Thus looking backward provides some insight into the structure of the future bits-and-atoms economy. It is probably not possible to estimate how much the maintenance of biological systems will add to GDP accounts around the world. But the point is to recognize that "off-book" maintenance will be occurring regardless of, and perhaps because of, the sophistication of the equipment used to synthesize and monitor complex biological systems.

Recall from Chapters 6–9 that, thanks to (1) automation, (2) the existence of relatively simple laboratory protocols, and (3) standardized biological parts, both standard services and cutting-edge innovation are already in the hands of high school and college students and workers with little or no formal educational training. Posit a world full of the fruits of mature biological engineering, wherein the fraction of GDP provided to developed economies by genetically modified organisms is potentially much larger than today's 2 percent: there *will* be substantial numbers of people involved in maintenance and repair of the relevant systems. By definition, those people *will* have all the training necessary to build, maintain, and operate those systems, and that activity *will* amount to significant economic value, even if it isn't accounted for directly. Most important, regardless of how and where they received that training, those workers are likely to employ their skills wherever they receive wages, as Edgerton notes: "Maintenance and repair are the most widespread forms of technical expertise . . . a great deal [of which] takes place outside the formal economy."[30] It *will not* be possible to keep track of maintenance activity in biotechnology because we do not maintain such records for any other sector of the economy: "Although central to our relationship with things, maintenance and repair are matters we would rather not think about. Mundane and infuriating, full of uncertainties, they are among the major annoyances surrounding things. The subject is left in the margins, often to marginal groups."[31]

In every technical area, in every developed country, perfectly functional goods are inevitably replaced by newer items, with the old going into service as lower-cost options next door or in the garage or, after being shipped, to less-developed countries. This is true for computers, cars, ships, airplanes, and cell phones, with maintenance and remanufacturing sectors emerging in India, Africa, South America, and Asia to support those items as they diffuse around the world.[32] Soon, even greater technical capability will be added to the existing service infrastructure via the eventual diffusion of low-cost Fab Labs, three-dimensional printers, and all the hardware necessary to build synthetic biological systems. Whatever the eventual shape of an economy based on distributed innovation, pervasive communication, and the fungibility of bits and atoms, that economy will rely on maintenance, repair, and widely dispersed technical expertise at least as much as the current economy does. Without an understanding of what has come before, we cannot understand what is yet to come.

Yet there are likely to be substantive differences between today and the eventual bits-and-atoms economy. Organisms producing fuels, materials, or structures will require tending—providing a new class of jobs—while the production itself will occur at the molecular level, largely out of sight of human handlers. More to the point, as production moves from cultured microbes to multicellular organisms and to plants grown in open fields, the structure of production begins to look more and more like nature itself.

Beyond the instruments that provide capabilities to synthesize, manipulate, and sequence DNA, no fundamentally new infrastructure is necessary to implement biological manufacturing. Instead, we must learn to use what already exists. In the world to come, the whole of biology constitutes our manufacturing infrastructure, and the whole of the terrestrial ecosystem sets the boundaries of an economy that must, somehow, be managed at a global level. We must learn to use the old. The vast majority of any "synthetic organism" built within at least the next decade will be composed primarily of old technology, inherited from billions of years of evolution.

The manufacturing "plant"—that takes instructions in the form of DNA and churns out materials, chemicals, and other functioning organisms—is at this time already far more advanced than we are presently capable of either understanding or utilizing to our own ends. It will be some time before our efforts at reprogramming the plant attain levels commensurate with the existing capabilities of the plant. And yet with efforts under way to expand the utility of genetic control circuitry and to expand both the genetic code and the diversity of amino acids used to implement protein designs, the plant is already growing in complexity and capability. But we still understand only very poorly the old system—the existing ecology—in which new organisms will live. Despite this uncertainty, modified biological systems will constitute an ever more-powerful manufacturing capability, with an ever-larger economic impact. Therefore, distributed biological innovation may have a much wider impact than open-source software and open innovation are having in the rest of the global economy.

Navigating Disequilibrium

"Equilibrium" is a word with many shades of meaning. It describes a balance of forces, outflow matching inflow, supply equaling demand, the rate forward being equivalent to the rate backward, stasis, maintaining equipoise. In none of these senses of the word is biological technology, or the growing markets and economy it supports, in equilibrium.

Biological technologies are becoming more sophisticated at a breakneck pace. Skills are proliferating internationally, providing additional skilled labor. There is constantly a greater demand for more food, better health care, and more fuel than the market can provide. Thus by our choices as consumers we point the way toward increased demand for greater innovation.

Where will new biological technologies and products come from? If William Baumol's David and Goliath theory of a division of labor in innovation holds for biological technologies, then the present structure of the bioeconomy is far out of equilibrium, dominated by large firms. Based on his historical and economic analysis, as costs fall and skills spread, we should expect tremendous innovation from entrepreneurs and small firms. If the ensuing proliferation of biological technologies enables distributed manufacturing, then the structure of whole sectors of the economy will have to find a new equilibrium.

Our accumulation of knowledge about biology and our ability to manipulate it are clearly proceeding at a startling pace. This book was written over several years, and within the final few months of writing came word of

- the first bacterial genome synthesized from scratch;
- the first data describing the energetic and carbon costs of widespread biofuels use;
- the first stable reprogramming of plant cells using artificial chromosomes;
- the first demonstration of embryonic stem cells derived from nuclear transfer;

- the first demonstration of pluripotent stem cells derived from somatic reprogramming;
- the first use of those cells to treat disease in an animal model;
- and the first use of autologous stem cells to produce a human organ used as a transplant.

What Should (and Should Not) Be Done

As we consider how to improve safety and security in the bioeconomy to come, it is worth paying attention to the experience of professionals attempting to improve the safety and security of the burgeoning information infrastructure. Excerpts from a short conversation between noted computer security experts Bruce Schneier and Markus Ranum provide some context:

> *Bruce Schneier: Throughout history and into the future, the one constant is human nature. There hasn't been a new crime invented in millennia. Fraud, theft, impersonation and counterfeiting are perennial problems that have been around since the beginning of society. During the last 10 years, these crimes have migrated into cyberspace, and over the next 10 they will migrate into whatever computing, communications and commerce platforms we're using.*
>
> *The nature of the attacks will be different: the targets, tactics and results. Security is both a trade-off and an arms race, a balance between attacker and defender, and changes in technology upset that balance. Technology might make one particular tactic more effective, or one particular security technology cheaper and more ubiquitous. Or a new emergent application might become a favored target.*
>
> *I don't see anything by 2017 that will fundamentally alter this.*
>
> *Marcus Ranum: I believe it's increasingly likely that we'll suffer catastrophic failures in critical infrastructure systems by 2017. It probably won't be terrorists that do it, though. . . . All the indicators point toward a system that is more complex, less well-understood and more interdependent. With infrastructure like that, who needs enemies?*
>
> *You're worried criminals will continue to penetrate into cyberspace, and I'm worried complexity, poor design and mismanagement will be there to meet them.*[33]

Threats to safety and security can arise from malfeasance and from poor design or implementation. While governments and other organizations will attempt to exert oversight and control, individuals and other organizations will resist, whether out of ideology, profit motive, or the simple joy of doing something completely new. To complicate the situation, new tools, new capabilities, and new vulnerabilities will constantly be introduced into our world.

How do we define biological safety and security in this world? The key to this endeavor is education. We cannot expect government or professional organizations to be the sole providers of security; it must also be in the hands of individuals. Developing an appreciation for potential benefit or harm is a critical component of public understanding, of public participation in both innovation and politics, and of policy choices.

Within the context of pervasive communication, distributed innovation, and the fungibility of bits and atoms, our physical and economic security depends on rapid and accurate perception of the world. This is the strongest argument for pursuing transparent technology development and for maximizing the flow of information. We should focus on three challenges:

(1) *We should resist the impulse to restrict research and the flow of information.* Ignorance will help no one in the event of an emergent threat and, given the pace and proliferation of biological technologies, the likelihood of threats will increase in coming years. One of the greatest threats we face is that potentially detrimental work will proceed while we sit on our hands. Science requires open communication, and a policy of voluntary ignorance serves only to guarantee economic and physical insecurity. If we are not ourselves pushing the boundaries of what is known about how pathogens work or ways to manipulate them, we are by definition at a disadvantage. Put simply, it will be much easier to keep track of what is in the wind if we don't have our heads in the sand.

(2) *The best way to keep apprised of the activities of both amateurs and professionals is to establish open networks.* We should give serious thought to government sponsorship of these networks. The open source–development community thrives on constant communication and plentiful free advice. Communication and collaboration is common practice for professional biology hackers, and it is already evident on the Web among amateur biology hackers.[34] This represents an opportunity to keep apprised of innovation in a distributed fashion. Anyone trying something new will require advice from peers and might advertise at least some portion of the results of his or her work. As is evident from the ready criticism leveled at miscreants in online forums frequented by software developers, people are not afraid to speak out when they feel the work of a particular person or group is substandard or threatens the public good. Thus our best potential defense against biological threats is to create and maintain open networks of

researchers at every level, thereby magnifying the number of eyes and ears keeping track of what is going on in the world.

(3) *Because human intelligence gathering is, alas, demonstrably inadequate for the task at hand, we should develop technology that enables pervasive environmental monitoring.* The best way to detect biological threats is using biology itself, in the form of genetically modified organisms. Unlike the production and deployment of chemical weapons or fissile materials, which can often be monitored with remote sensing technologies such as in situ video surveillance, as well as aerial and satellite reconnaissance, the initial indication of biological threats may be only a few cells or molecules. This small quantity may already be a lethal dose and can be very hard to detect using devices solely based on chemistry or physics. As an alternative, "surveillance bugs"—engineered organisms—distributed in the environment could transduce small amounts of cells or molecules into signals measurable by remote sensing. The organisms might be modified to reproduce in the presence of certain signals, to change their schooling or flocking behavior, or to alter their physical appearance. Candidate "detector platforms" span the range of bacteria, insects, plants, and animals. Transgenic zebrafish, nematodes, and the mustard weed *Arabidopsis thaliana* have already been produced for this purpose.[35] The arsenic biosensor developed by the 2006 iGEM team from Edinburgh (see Chapter 7) is a clear indication that this approach can provide needed capabilities. Engineered organisms might either be integrated as sensors into an enclosed device, or might be released into the wild; the later choice would face scrutiny similar to any other genetically modified crop or animal.

None of the above recommended goals will be trivial to accomplish. Considerable sums have been spent over the last five decades to understand biological systems at the molecular level, much of this in the name of defeating infectious disease. While this effort has produced remarkable advances in diagnosing and treating disease, we should now redouble our efforts.

Over the last several years there have been various public calls for "a new Manhattan Project" to develop countermeasures against both natural and artificial biological threats.[36] As I describe in Chapter 9, we absolutely require new technology to combat both natural and artificial pathogens, but the Manhattan Project is decidedly the wrong model for an effort to increase biological security.

Previous governmental efforts to rapidly develop technology, such as the Manhattan and Apollo Projects, were predominantly closed, arguably with good reason at the time. But we live in a different era and should consider an open effort that takes advantage of preexisting research and development networks. This strategy may result in more robust, sustainable, distributed security and economic benefits. Note also that although both the Manhattan and Apollo Projects both were closed and centrally coordinated, they were very different in structure. The Apollo Project took place in the public eye, with failures plainly writ in smoke and debris in the sky. The Manhattan Project, in contrast, took place behind barbed wire and was so secret that very few people within the U.S. government and military knew of its existence. A secret project is not the ideal model for research that is explicitly aimed at understanding how to modify biological systems. Above all else, let us insist that this work happens in the light, subject to the scrutiny of all who choose to examine it.

Beyond providing an innate intelligence-gathering capability, open and distributed networks of researchers and innovators would provide a flexible and robust workforce for developing technology. This resource could be employed in rapid reaction to emerging threats and the development of a response that might include assembling novel compounds or organisms. The rudiments of a response system were demonstrated during the recent SARS outbreak, but much more is required.[37]

Some may consider several decades of experience with open-source software insufficient as an organizational model to serve as a basis for a response to biological threats. The best model may in fact be found in the history of biology as an academic discipline. In order to bring the focus of an Apollo Project to the task at hand, the traditions of open discourse among academics and the sharing of reagents and biological stocks might be strengthened and adapted. Hoarding of results or materials should be strongly discouraged, and in fact sharing information and stocks should be required. It may be prudent to write down these guidelines in documents with legal standing, if only to give added weight to peer pressure. To be sure, legal guidelines constructed in this way might be viewed as a form of self-regulation, but this would be in the context of open markets. Open access to information and technology might serve to suppress the black markets that emerge under regulation from above. These agreements could be structured so that voluntary participation would provide ready access to information or reagents otherwise difficult to procure, thereby encouraging participation but not outlawing the activities of those who choose to remain independent. The ability to compete in the market might therefore be enhanced through cooperation and access to information, whereas scientific and technological development outside the community might be less efficient and thereby inhibited.

New or existing foundations might take these agreements in hand to provide coordination. There is already some structure of this sort extant in the biological community, with organizations such as the American Cancer Society, the Wellcome Trust, and the Bill and Melinda Gates Foundation, among many others, providing funding for meetings, journals, physical infrastructure, and particular directions of research.

Finally, the best argument for encouraging the development of open biology in amateur, academic, and industrial contexts is that the resulting community will be better able to spot and respond to errors. This goal is nowhere more important than in the burgeoning enterprise of manipulating life at the genetic level. Creating international networks that coordinate an open biology may be the most important step we can take to improve our security in the coming decades.

THE HISTORY of any given technology is extraordinarily complex. Individuals come and go, sometimes playing crucial roles both via their inventions and via their role in commercial innovation.

Technologies can take many decades to penetrate an economy, particularly in the context of preexisting investment in alternatives or of cultural resistance. Biological technologies may be no different, even in circumstances where a biological process is particularly efficient or inexpensive.

The benefits of exploring this frontier are both exceptionally promising and, even after three decades of developing recombinant DNA technologies, largely and frustratingly elusive. But the work will continue. The potential benefits are too scientifically, politically, and economically enticing for humans to resist. Individuals and governments alike are fascinated by the possibility of improved crop yields, increased meat production, plentiful biofuels, and improved human health through new vaccines and replacement tissues.

This book has told a story about learning to build predictable biological systems. Just as Mouillard realized over a century ago that humans would get off the ground only by learning to fly, rather than by slapping an engine and some winglike parts together in the hopes that flight would ensue, so are we today embarking on learning how biological bits and pieces fit together to make functioning objects. What Mouillard, the Wright Brothers, Octave Chanute, and all the other early aviation pioneers did not foresee was the effect their work would have on our society and culture as a whole. Similarly, neither those innovators nor the citizens of any one country were able to exert substantial control over the development of the technology for very long.

Over the early decades of the twentieth century, governments, military organizations, and corporations all pushed aviation in particular directions to suit their needs. What began as an intellectual challenge rapidly evolved through a tactical military advantage into an economic revolution that modern economies cannot survive without. Development within the aeronautics industry has at some times proceeded faster than at others, but air transport is now a critical link in the economies of countries containing most of the world's population. Whether this situation represents ideal policy, or wise apportionment of economic resources, is beside the point. Here we are; this is the world we live in. What follows comes from today.

This book has also explored historical examples that provide clear models for maximizing technological innovation. Arguments that we should eschew maximizing that innovation, either through regulated access to technology or through constraining governmental and industrial funding, will leave us poorly prepared to deal even with natural threats, let alone the inevitable emergence of artificial threats. Even in the face of any attempt to regulate access, as costs fall and skills spread, adoption of biological technologies will be widespread.

The broader revolution of distributed innovation, pervasive communication, and the fungibility of bits and atoms makes regulating access all the more likely to fail. The direct relevance of this revolution to biological technologies is that even if we attempted to regulate the parts for DNA synthesizers or other equipment, rapid prototyping equipment and three-dimensional printers could be used to reproduce those components. In addition, prohibition is generally short lived and ineffective. Those arguing for attempting to improve safety and security through regulation and restriction must demonstrate successful examples of such policies within market economies. Front-end regulation will hinder the development of a thriving industry driven and supported by entrepreneurs and thereby engender a world that is less safe.

Our society is only just beginning to struggle with all the social and technical questions that arise from a fundamental transformation of the economy. History holds many lessons for those of us involved in creating new tools and new organisms and in trying to safely integrate these new technologies into our complex socioeconomic system. Alas, history also fails to provide examples of any other technological system as powerful as rational engineering of biology, and we have precious little guidance concerning how our socioeconomic system might be changed in the years to come. We can attempt only to minimize our mistakes and rapidly correct those we and others do make. In a few years the path might become clearer, after the trailblazers try many approaches, in essence providing early experimental data defining the future shape of our society and economy.

The coming bioeconomy will be based on fundamentally less-expensive and more-distributed technologies than those that shaped the course of the twentieth century. Our choices about how to develop biological technologies will determine the pace and effectiveness of innovation as a public good. As with the rest of the natural and human-built world, the development of this system is entirely in human hands.

Afterword

THE HARDEST part of writing this book was keeping pace with changes in biological technologies. It was like trying to maintain one's footing on shifting sands during an earthquake while a hurricane comes ashore.

The story about synthetic biological systems started with Michael Elowitz, Tim Gardner, and Drew Endy in Chapter 4, and touched on iGEM and a few other individuals along the way, but I was unable to discuss scores of excellent published examples of synthetic biological systems that display predictable behaviors. This omission occurred simply because everyone in the field is running so fast that it is impossible to keep up. Students, as usual, are leading the way.

In the 2008 iGEM competition, teams presented projects on everything from bacterial electricity production, to building new engineering tools, to synthetic vaccines.[1] Those vaccines, developed by the grand prize winner Team Slovenia—composed of undergraduates—were generated and then tested in mice within a few months, demonstrating both the utility of synthesis as a rapid response technology (see Chapter 9) and the broad spread of biological technologies around the globe.

We may be faced with the need to deploy synthetic vaccines very soon. The surprise pathogen of 2009 is indeed influenza, but not the strain everyone was keeping an eye on, and it did not show up where everyone was looking. In 2009, H1N1 was first identified in Mexico and rapidly spread worldwide. While the strain can cause severe illness and death, it is not presently as bad as had been initially feared. Vaccine production has already started, but we are clearly in the situation described in Chapter 9; viral evolution may outpace our existing vaccine production technology. Synthetic vaccines, including DNA vaccines, are already under discussion as a means to escape this trap.

Gene synthesis has also been used to recreate another pathogen from scratch. In late 2008, Ralph Baric and his collaborators described a method for rebuilding the SARS coronavirus, a technology that is required for understanding the biology and evolution of the virus.[2] The technology also, obviously, increases the risk that the pathogen could be intentionally released. Again we are stuck with the inherent dual-use nature of biological technologies. Most human technologies are not dangerous in and of themselves; humans make a technology dangerous through intent and use. Biological technologies developed by nature, however, can be dangerous to humans without any intervention or invention on our part, and only through effort and ingenuity will we reduce the threat of pandemic influenza, SARS, or malaria.

The broader message in this book is that biological technologies are beginning to change both our economy and our interaction with nature in new ways. The global acreage of genetically modified (GM) crops continues to grow at a very steady rate, and those crops are put to new uses in the economy every day. One critical question I avoided in the discussion of these crops is the extent to which GM provides an advantage over unmodified plants. With more than ten years of field and market experience with these crops in Asia and North and South America, the answer would appear to be yes. Farmers who have the choice to plant GM crops often do so, and presumably they make that choice because it provides them a benefit. But public conversation remains highly polarized. The Union of Concerned Scientists recently released a review of published studies of GM crop yields in which the author claimed to "debunk" the idea that genetic modification will "play a significant role in increasing food production."[3] The Biotechnology Industry Organization responded with a press release claiming to "debunk" the original debunking.[4] The debate continues.

Similarly, the conversation about using crops to make fuels remains passionate, although the large-scale manufacturing of biofuels has apparently been delayed somewhat by a global economic slowdown. That slowdown has caused large fluctuations in the price of petroleum, undermining the economic impetus to produce biofuels. But as the global economy recovers, the price of oil, and the price of food, will undoubtedly rise again. Hopefully, improved biological technologies will ease, rather than exacerbate, the competition between food and fuel.

The ongoing debates over just which biofuel is best, which production pathway requires the least energy and emits the least carbon dioxide, and which feedstock is most economically viable and environmentally benign, all reemphasize the central theme of the book. Biology is technology. As with any other technology our evaluation—and *valuation*—of biology will change with perception, new data, and irrepressible, inevitable innovation.

Updates to the story told in this book can be found at: www.biologyistechnology.com.

Notes

1. What Is Biology?

1. Bureau of Economic Analysis, Industry Economic Accounts, "Value added by industry as a percentage of gross domestic product," www.bea.gov/industry/gpotables/gpo_action.cfm?anon=78432&table_id=22073&format_type=0.
2. See the CIA *World Factbook,* www.cia.gov/library/publications/the-world -factbook/index.html.
3. Wikipedia, http://en.wikipedia.org/wiki/Market (accessed 20 September 2008).
4. J. Tierney, "An early environmentalist, embracing new 'heresies,'" *New York Times,* 27 February 2007.

2. Building with Biological Parts

1. SwayStudio, "Purpose," www.swaystudio.com/honda_movie.html. The press material for the spot describes the plastic pieces as "Mega Bloks," which appear to be a knockoff of products from the famous Danish company. No one could possibly mistake the shape, however.
2. H. Liu, R. Ramnarayanan, and B. E. Logan, *Production of electricity during wastewater treatment using a single chamber microbial fuel cell,* Environmental Science & Technology 38, no. 7 (2004): 2281–2285.
3. H. Ceremonie et al., *Isolation of lightning-competent soil bacteria.* Appl Environ Microbiol, 70, no. 10 (2004): 6342–6346.
4. J. W. Beaber, B. Hochhut, and M. K. Waldor, *SOS response promotes horizontal dissemination of antibiotic resistance genes,* Nature 427, no. 6969 (2004): 72–74.

5. N. V. Fedoroff, *Agriculture: Prehistoric GM corn,* Science 302, no. 5648 (2003): 1158–1159.
6. J. Kaiser, *Gene therapy: Seeking the cause of induced leukemias in X-SCID trial,* Science 299, no. 5606 (2003): 495; J. L. Fox, *US authorities uphold suspension of SCID gene therapy,* Nat Biotechnol 21, no. 3 (2003): 217.
7. N.-B. Woods et al., *Gene therapy: Therapeutic gene causing lymphoma,* Nature 440, no. 7088 (2006): 1123.

3. Learning to Fly (or Yeast, Geese, and 747s)

1. A. Kvist et al., *Carrying large fuel loads during sustained bird flight is cheaper than expected,* Nature 413, no. 6857 (2001): 730–732.
2. H. Weimerskirch et al., *Energy saving in flight formation.* Nature 413, no. 6857 (2001): 697–698.
3. R. H. MacArthur and E. O. Wilson, *The Theory of Island Biogeography,* Monographs in Population Biology, 1st series (Princeton, N.J.: Princeton University Press, 1967), xi, 203.
4. L.-P. Mouillard, *L'empire de l'air: Essai d'ornithologie appliquée a l'aviation* (G. Masson: Paris, 1881).
5. J. Bevo-Higgins, ed., *The Chanute-Mouillard Correspondence, April 16, 1890, to May 20, 1897* (San Francisco: E. L. Sterne, 1962) p. 51. The expense and difficulty involved in processing aluminum would prevent its widespread use in aircraft until just before the Second World War. During the war years, more airplanes were built than in the previous four decades combined, in part because hydroelectric dams constructed in the United States and Europe in the 1920s and 1930s provided copious power for the electrolytic purification of aluminum. See www.world-aluminium.org/production/smelting/index.html and www.wpafb.af.mil/museum/history/wwii/aaf/aaf30.htm.
6. Bevo-Higgins, *Chanute-Mouillard Correspondence.*
7. The National Air and Space Museum contains a replica of a Lilienthal glider. A photo and description are available at www.nasm.edu/nasm/aero/aircraft/lilienth.htm.
8. M. D. Ardema, J. Mach, and W. J. Adams, "John Joseph Montgomery 1883 glider," 11 May 1996, American Society of Mechanical Engineers. Various sources place this flight in 1894 rather than 1893 and suggest that the 1893 test was a failed attempt at powered flight using flapping wings.
9. M. W. McFarland, ed., *The Papers of Wilbur and Orville Wright, Including the Chanute-Wright Letters and Other Papers of Octave Chanute* (New York: McGraw-Hill, 1953).
10. Ibid.
11. There has been some controversy during the last one hundred years surrounding the assertion that the Frenchman Clément Ader made the first powered flight as early as 1890. However, based on its design, Mouillard dismissed Ader's airplane as a "phantasmagoria" (Bevo-Higgins, *Chanute-Mouillard Correspondence,* 54). The Wrights similarly concluded it could not physically

fly and claimed in correspondence with Chanute to have testimony to that effect from a French army officer present at a failed test of Ader's aircraft (McFarland, *Papers of Wilbur and Orville Wright*, 952).

12. McFarland, *Papers of Wilbur and Orville Wright*.
13. Ibid.
14. K. Sabbagh, *21st Century Jet: The Making and Marketing of the Boeing 777* (New York: Scribner, 1996), 336.
15. "Help! There's nobody in the cockpit," *Economist*, 21 December 2002.
16. D. Voet and J. G. Voet, *Biochemistry*, 2nd ed. (New York: John Wiley & Sons, 1995); I. Levine, *Physical Chemistry* (New York: McGraw-Hill Companies, 1988).
17. Roche Applied Science publishes the most visually impressive versions of these charts. They are available at www.expasy.org/cgi-bin/search-biochem-index.

4. The Second Coming of Synthetic Biology

1. See "Friedrich-Wohler: Aluminum-and-urea-papers," *Encyclopedia Britannica*, www.britannica.com/EBchecked/topic/646422/Friedrich-Wohler/259994/Aluminum-and-urea-papers; "Wöhler to Berzelius (1828)," https://webspace .yale.edu/chem125_fo6/125/history99/4RadicalsTypes/UreaLetter1828.html.
2. S. Toby, *Acid test finally wiped out vitalism, and yet,* Nature 408, no. 6814 (2000): 767–767.
3. A. J. Rocke, *The Quiet Revolution: Hermann Kolbe and the Science of Organic Chemistry* (Berkeley: University of California Press, 1993), 60.
4. E. F. Keller, *Making Sense of Life: Explaining Biological Development with Models, Metaphors, and Machines* (Cambridge, Mass.: Harvard University Press, 2002), 388; quotation, 18.
5. Ibid., 28.
6. Ibid., 31.
7. Ibid.
8. Ibid., 86.
9. W. Szybalski, "In Vivo and in Vitro Initiation of Transcription," in A. Kohn and A. Shatkay, eds., *Control of Gene Expression* (New York: Plenum Press, 1974).
10. W. Szybalski and A. Skalka, *Nobel prizes and restriction enzymes,* Gene 4, no. 3 (1978): 181–182.
11. S. A. Benner and A. M. Sismour, *Synthetic biology,* Nat Rev Genet, 6, no. 7 (2005): 533–543.
12. The best cartoon description of biology is quite literally a cartoon. Larry Gonick and Mark Wheelis's *Cartoon Guide to Genetics* (New York: Barnes and Noble, 1983) is a masterful description of molecular biology written at a level that is easily accessible to nonscientists. Dog-eared copies can also be found on the bookshelves of many professionals.
13. O. T. Avery, C. M. MacLeod, and M. McCarty, *Studies on the chemical nature of the substance inducing transformation of pneumococcal types: Induction of*

transformation by desoxyribonucleic acid fraction isolated from Pneumococcus Type III, J Exp Med 79, no. 2 (1944): 137.

14. F. Crick, "On Protein Synthesis," at Symposium of the Society of Experimental Biology 12 (1957): 138–163.

15. T. S. Gardner, C. R. Cantor, and J. J. Collins, *Construction of a genetic toggle switch in* Escherichia coli, Nature 403, no. 6767 (2000): 339–342.

16. D. Voet and J. G. Voet, *Biochemistry,* 2nd ed. (New York: John Wiley & Sons, 1995).

17. M. B. Elowitz and S. Leibler, *A synthetic oscillatory network of transcriptional regulators,* Nature 403, no. 6767 (2000): 335–338.

18. D. Endy et al., *Computation, prediction, and experimental tests of fitness for bacteriophage T7 mutants with permuted genomes,* Proc Natl Acad Sci U.S.A. 97 (2000): 5375–5380.

19. Ibid.

20. L. Y. Chan, S. Kosuri, and D. Endy, *Refactoring bacteriophage T7,* Mol Syst Biol 1 (2005): 0018.

5. A Future History of Biological Engineering

1. See "Schools participating in iGEM 2006," http://parts2.mit.edu/wiki/index .php/Schools_Participating_in_iGEM_2006.

2. See "Registration FAQ," http://parts2.mit.edu/wiki/index.php/Registration_FAQ.

3. S. Berkun, *The Myths of Innovation* (Sebastopol, Calif.: O'Reilly, 2007).

4. J. Newcomb, R. Carlson, and S. Aldrich, *Genome Synthesis and Design Futures: Implications for the U.S. Economy* (Cambridge, Mass.: Bio Economic Research Associates, 2007), 35. W. B. Arthur and W. Polak, "The evolution of technology within a simple computer model," *Complexity,* 2006 11(5): 23–31.

5. M. R. Darby and L. G. Zucker, "Grilichesian breakthroughs: Inventions of methods of inventing and firm entry in nanotechnology" (Working Paper 9825, National Bureau of Economic Research, Cambridge, Mass., July 2003). Griliches, Zvi, "Hybrid Corn: An Exploration in the Economics of Technological Change," *Econometrica,* October 1957, 25(4): 501–522.

6. Newcomb, Carlson, and Aldrich, *Genome Synthesis and Design Futures.*

7. P. Brimlow, "The silent boom," *Forbes,* 7 July 1997.

8. Newcomb, Carlson, and Aldrich, *Genome Synthesis and Design Futures,* 57.

6. The Pace of Change in Biological Technologies

1. Discovery DNA Explorer Kit, #691907, www.discovery.com.

2. J. Cello, A. V. Paul, and E. Wimmer, *Chemical synthesis of poliovirus cDNA: Generation of infectious virus in the absence of natural template,* Science 297, no. 5583 (2002): 1016–1018.

3. E. Pilkington, "I am creating artificial life, declares US gene pioneer," *Guardian,* 6 October 2007; D. G. Gibson et al., *Complete chemical synthesis, assembly, and cloning of a* Mycoplasma genitalium *genome,* Science 319, no. 5867 (2008): 1215. My sources for Figure 6 include: Khorana, H. G., *Total synthesis of a gene.*

Science, 1979. 203(4381): 614–625; Mandecki, W., et al., *A totally synthetic plasmid for general cloning, gene expression and mutagenesis in Escherichia coli.* Gene, 1990. 94(1): 103–107; Stemmer, W. P., et al., *Single-step assembly of a gene and entire plasmid from large numbers of oligodeoxyribonucleotides.* Gene, 1995. 164(1): 49–53; Cello, J., A. V. Paul, and E. Wimmer, *Chemical synthesis of poliovirus cDNA: generation of infectious virus in the absence of natural template.* Science, 2002. 297(5583): 1016–1018; Tian, J., et al., *Accurate multiplex gene synthesis from programmable DNA microchips.* Nature, 2004. 432(7020): 1050–1054; Kodumal, S. J., et al., *Total synthesis of long DNA sequences: synthesis of a contiguous 32-kb polyketide synthase gene cluster.* Proc Natl Acad Sci U S A, 2004. 101(44): 15573–15578; Gibson, D. G., et al., *One-step assembly in yeast of 25 overlapping DNA fragments to form a complete synthetic Mycoplasma genitalium genome.* Proc Natl Acad Sci U S A, 2008. 105(51): 20404–20409.

4. G. Moore, *Cramming more components onto integrated circuits,* Electronics 38, no. 8 (1965):114–117.

5. Ibid., 114.

6. M. Ronaghi, *Pyrosequencing sheds light on DNA sequencing,* Genome Research 11, no. 1 (2001): 3–11.

7. "Nobel laureate James Watson received personal genome," 13 August 2007, available at www.sciencedaily.com/releases/2007/05/070531180739.htm.

8. F. S. Collins, M. Morgan, and A. Patrinos, *The Human Genome Project: Lessons from large-scale biology,* Science 300, no. 5617 (2003): 286–290.

9. S.-C. J. Chen, "Under new management," *The Economist,* April 2, 2009.

10. Genome Technology, editorial, April 2001. (magazine)

11. E. S. Lander et al., *Initial sequencing and analysis of the human genome,* Nature 409, no. 6822 (2001): 860–921.

12. R. F. Service, *Gene sequencing: The race for the $1000 genome,* Science 311, no. 5767 (2006): 1544–1546.

13. D. S. Kong et al., *Parallel gene synthesis in a microfluidic device,* Nucl Acids Res 35, no. 8 (2007): e61.

14. P. A. Carr et al., *Protein-mediated error correction for de novo DNA synthesis,* Nucl Acids Res 32, no. 20 (2004): e162.

15. Brian Baynes, Codon Devices, personal communication.

16. I. Braslavsky et al., *Sequence information can be obtained from single DNA molecules,* Proc Natl Acad Sci U.S.A. 100, no. 7 (2003): 3960–3964.

17. See, for example, Garner Lab, http://innovation.swmed.edu/Instrumentation/mermade_oligonucleotide_synthesi.htm.

18. DNA Synthesis Panel, International Conference on Synthetic Biology 2.0, University of California, Berkeley.

19. Map of commercial gene foundaries, http://synthesis.typepad.com/synthesis/2005/07/global_distribu.html.

20. "A genome shop near you," *Wired,* December 2005, www.wired.com/wired/archive/13.12/start.html?pg=16; J. Newcomb, R. Carlson, and S. Aldrich, *Genome Synthesis and Design Futures: Implications for the U.S. Economy* (Cambridge, Mass.: Bio Economic Research Associates, 2007), 16.

21. M. Garfinkel et al., *Synthetic Genomics: Options for Governance* (Rockville, Md.: J. Craig Venter Institute, in collaboration with CSIS, MIT, 2007).

22. H. Bügl et al., *DNA synthesis and biological security,* Nat Biotech 25, no. 6 (2007): 627.

7. The International Genetically Engineered Machines Competition

1. P. Di Martino et al., *Indole can act as an extracellular signal to regulate biofilm formation of* Escherichia coli *and other indole-producing bacteria,* Canadian Journal of Microbiology 49, no. 7 (2003): 443; J. Lee, A. Jayaraman, and T. K. Wood, *Indole is an inter-species biofilm signal mediated by SdiA,* BMC Microbiol 7 (2007): 42.

2. Shetty, Reshma, "Applying engineering principles to the design and construction of transcriptional devices," Thesis (PhD), MIT, Biological Engineering Division, 2008; Regarding the educational experience of the undergraduates: Reshma Shetty, personal comm.

3. C. Mead and L. Conway, *Introduction to VLSI Systems* (Reading, Mass.: Addison-Wesley, 1980).

4. G. Smith, "Unsung innovators: Lynn Conway and Carver Mead," *Computerworld,* 3 December 2007.

5. "Thoughts on the biology/EECS relationship," Tom Knight, 23 January 2003, additions 26 May 2005, www.eecs.mit.edu/bioeecs/Knight_essay.html.

6. Ibid.

7. "2005 supplement," at "Parts, devices, systems & engineering theory," http://parts.mit.edu/wiki/index.php/Parts%2C_Devices%2C_Systems_%26 _Engineering_Theory.

8. For descriptions of the IAP projects, see "IAP 2003 and IAP 2004 Projects," http://parts.mit.edu/projects/index.cgi.

9. Reshma Shetty, personal communication.

10. For a firsthand account of iGEM 2004, see A. M. Campbell, *Meeting report: Synthetic biology jamboree for undergraduates,* Cell Biology Education 4 (2005): 19–13.

11. See also D. Endy, *Foundations for engineering biology,* Nature 438, no. 7067 (2005): 449.

12. See "SBC 2004—UT Austin," http://parts.mit.edu/r/parts/htdocs/SBC04/austin .cgi.

13. A. Levskaya et al., *Synthetic biology: Engineering* Escherichia coli *to see light,* Nature 438, no. 7067 (2005): 441.

14. Christopher Voigt, personal communication.

15. In order to be correctly assembled by the cell, BBa_I15010 requires the presence of two other parts, genes that code for proteins necessary to preprocess metabolites into components of the photoreceptor complex. These genes are defined in the specification for the phototransducer but play no role in the real-time function of the transducer.

16. Levskaya, *Synthetic biology,* 441.

17. See Campbell, *Meeting report.*

18. See "iGEM 2005," http://parts.mit.edu/wiki/index.php/Igem_2005.
19. E. Check, *Synthetic biology: Designs on life,* Nature 438, no. 7067 (2005): 417.
20. See "iGEM 2005."
21. Check, *Synthetic biology,* 417.
22. See www.igem2006.com/jamboree.htm. For a map of participating schools, see www.igem2006.com/meet.htm.
23. See "iGEM 2006 Jamboree results," www.igem2006.com/results.htm.
24. See www.igem2006.com/presentations.htm.
25. See IET Synthetic Biology 1, issues 1–2 (June 2007): 1–90, http://scitation .aip.org/dbt/dbt.jsp?KEY=ISBEBU&Volume=CURVOL&Issue=CURISS.
26. See "Engineered human cells: Say no to sepsis," http://parts.mit.edu/wiki/index .php/ Ljubljana%2C_Slovenia_2006.
27. See *Deaths: Final data for 2005,* National Vital Statistics Report 56, no. 10, www.cdc.gov/nchs/data/nvsr/nvsr56/nvsr56_10.pdf; F. Jaimes, *A literature re- view of the epidemiology of sepsis in Latin America,* Rev Panam Salud Pub- lica 18, no. 3 (2005): 163–171.
28. See "Programmed differentiation of mouse embryonic stem cells using artifi- cial signaling pathways," http://parts.mit.edu/wiki/index.php/Princeton:Project _Summary.
29. Ron Weiss, personal communication.
30. See "Towards an assessment of the socioeconomic impact of arsenic poison- ing in Bangladesh," World Health Organization, 2000, www.who.int/water _sanitation_health/dwq/arsenic2/en/index.html.
31. Chris French, personal communication.
32. See iGEM 2007 wiki, http://parts.mit.edu/igem07/index.php/Main_Page.
33. See "Towards self-differentiated bacterial assembly line," http://parts.mit.edu/ igem07/index.php/Peking_The_Projects.
34. See "The SMB: Synthetic multicellular bacterium," http://parts.mit.edu/igem07/ index.php/Paris.
35. See "USTC iGEM 2007," http://parts.mit.edu/igem07/index.php/USTC.
36. See "Location, location, location: Directing biology through synthetic assem- blies and organelles," http://parts.mit.edu/igem07/index.php/UCSF.
37. See "Bactoblood," http://parts.mit.edu/igem07/index.php/Berkeley_UC.
38. See "Virotrap: A synthetic biology approach against HIV," http://parts.mit.edu/ igem07/index.php/Ljubljana.
39. See, for example, "Moore's Law," www.intel.com/technology/mooreslaw/index .htm?iid=intel_tl+moores_law.

8. Reprogramming Cells and Building Genomes

1. J. D. Keasling, "Synthetic biology in pursuit of low-cost, effective, anti-malarial drugs" (presentation at the Institute for Systems Biology Seventh Annual Inter- national Symposium, Seattle, Wash., 21 April 2008).
2. M. Ettling et al., *Economic impact of malaria in Malawian households,* Trop Med Parasitol 45, 1 (1994): 74–79; J. M. Chuma, M. Thiede, and C. S. Molyneux, *Rethinking the economic costs of malaria at the house-*

hold level: Evidence from applying a new analytical framework in rural Kenya, Malar J 5 (2006): 76. See also "The World Health Organization, Health and Environment Linkages Initiative: Human toll and economic costs of malaria," www.who.int/heli/risks/vectors/malariacontrol/en/index1.html.

3. Sandi Doughton, "Gates Foundation tackles a giant that preys on Africa's children," *Seattle Times,* 6 February 2008.

4. J. L. Gallup and J. D. Sachs, *The economic burden of malaria,* American Journal of Tropical Medicine and Hygiene 64, suppl. 1 (2001): 85.

5. See "Economic costs of malaria are many times higher than previously estimated," 25 April 2000, www.malaria.org/news239.html.

6. Ibid.

7. Bill Rau, "Too Poor to be Sick: Linkages Between Agriculture and Health," UN/FAO, 2006.

8. For a brief history, see N. J. White, *Qinghaosu (artemisinin): The price of success,* Science 320, no. 5874 (2008): 330–334.

9. Jay Keasling, "Synthetic biology in pursuit of low-cost, effective, anti-malarial drugs."

10. S. Connor, "Malaria: A miracle in the making offers hope to millions worldwide," *Independent,* 4 June 2008.

11. V. J. Martin et al., *Engineering a mevalonate pathway in* Escherichia coli *for production of terpenoids,* Nat Biotechnol 21, no. 7 (2003): 796–802.

12. Keasling, "Synthetic biology in pursuit of low-cost, effective, anti-malarial drugs."

13. Ibid.

14. S. T. Payne et al., *Eau d'*E coli: *Rapid prototyping of a genetically-encoded olfactory reporter system,* 2008, MIT.

15. On the billionfold increase in yield, see Jay Keasling, UC Berkeley Faculty Forum on the Energy Biosciences Institute, 8 March 2007, http://webcast.berkeley.edu/event_details.php?webcastid=19207&p=1&ipp=1000&category=; on the necessary increase in yield to make production economically viable, see Keasling, "Synthetic biology in pursuit of low-cost, effective, anti-malarial drugs."

16. Ibid.

17. M. H. Serres, S. Goswami, and M. Riley, *GenProtEC: An updated and improved analysis of functions of* Escherichia coli *K-12 proteins,* Nucleic Acids Res 32, database issue (2004): D300–D302.

18. For genome length, see G. Posfai et al., *Emergent properties of reduced-genome* Escherichia coli, Science 312, no. 5776 (2006): 1044–1046; for Clean Genome *E. coli,* see Scarab Genomics, "Products," www.scarabgenomics.com.

19. Posfai et al., *Emergent properties of reduced-genome* Escherichia coli.

20. Tom Knight, Second International Conference on Synthetic Biology (University of California, Berkeley, May 2006).

21. Tom Knight, "Thoughts on the biology/EECS relationship," 23 January 2003, additions 26 May 2005, www.eecs.mit.edu/bioeecs/Knight_essay.html.

22. D. G. Gibson et al., *Complete chemical synthesis, assembly, and cloning of a* Mycoplasma genitalium *genome,* Science 319, no. 5867 (2008): 1215.

23. Clyde Hutchison, Fourth International Conference on Synthetic Biology (Hong Kong University of Science and Technology, 10 October 2008).

24. Ibid.
25. D. G. Gibson et al., *One-step assembly in yeast of 25 overlapping DNA fragments to form a complete synthetic* Mycoplasma genitalium *genome,* Proc Natl Acad Sci U.S.A. 105, no. 51 (2008): 20404–20409.
26. A. Pollack, "Scientists take new step toward man-made life," *New York Times,* 24 January 2008.
27. J. I. Glass et al., *Essential genes of a minimal bacterium,* Proc Natl Acad Sci U.S.A. 103, no. 2 (2006): 425–430.
28. See Craig Venter, "Creating life in a lab using DNA," *Telegraph,* 16 October, 2007.
29. Z. Shao and H. Zhao, *DNA assembler, an in vivo genetic method for rapid construction of biochemical pathways,* Nucleic Acids Res 37, no. 2 (2009): e16.

9. The Promise and Peril of Biological Technologies

1. M. Enami et al., *Introduction of site-specific mutations into the genome of influenza virus,* PNAS 87, no. 10 (1990): 3802–3805.
2. G. Neumann et al., *Generation of influenza A viruses entirely from cloned cDNAs,* PNAS 96, no. 16 (1999): 9345–9350.
3. G. Neumann et al., *An improved reverse genetics system for influenza A virus generation and its implications for vaccine production,* PNAS 102, no. 46 (2005): 16825–16829.
4. *The 1918 flu virus is resurrected,* Nature 437, no. 7060 (2005): 794.
5. J. C. Kash et al., *Genomic analysis of increased host immune and cell death responses induced by 1918 influenza virus,* Nature 443, no. 7111 (2006): 578.
6. D. Kobasa et al., *Aberrant innate immune response in lethal infection of macaques with the 1918 influenza virus,* Nature 445, no. 7125 (2007): 319.
7. Y.-M. Loo and M. Gale, *Influenza: Fatal immunity and the 1918 virus,* Nature 445, no. 7125 (2007): 267.
8. Repeated requests to Jefferey Taubenberger for clarification on the issue of ease of influenza viral reconstruction went unmet, even when mediated by one of his collaborators. While somewhat frustrating from my perspective, his reticence to discuss the issue is entirely understandable on a personal and professional level and may well be part of official U.S. government or military strategy to contain proliferation. But more-open discussion about how hard the process is would benefit both policy debates and basic science.
9. CDC, "Influenza activity—United States and worldwide, 2007–08 season," *Morbidity and Mortality Weekly Report,* 26 June 2008.
10. *The 1918 flu virus is resurrected.*
11. M. Garfinkel et al., *Synthetic Genomics: Options for Governance* (Rockville, Md.: J. Craig Venter Institute, in collaboration with CSIS, MIT, 2007), 12.
12. *The 1918 flu virus is resurrected.*
13. C. Sheridan, *Next generation flu vaccine boosted by Chiron debacle,* Nat Biotechnol 22, no. 12 (2004): 1487–1488.
14. On government participation in the market, see *National Strategy for Pandemic Flu,* Homeland Security Council, the White House, 2005, at www.pandemicflu.gov; on existing technologies' ability to produce sufficient num-

bers of doses, see M. T. Osterholm, *Preparing for the next pandemic,* N Engl J Med 352, no. 18 (2005): 1839–1842; Statement of Marcia Crosse, Director, Health Care, testimony before the Subcommittee On Health, Committee on Energy and Commerce, House of Representatives, *Influenza Pandemic: Challenges Remain in Preparedness* (Washington D.C.: Government Accountability Office, 2005); D. Butler, *Bird flu vaccine not up to scratch,* Nature News, 10 August 2005.

15. *BBC News,* "Gloomy estimate of bird flu costs," http://news.bbc.co.uk/1/hi/world/asia-pacific/4414668.stm; World Bank, http://siteresources.worldbank.org/INTEAPHALFYEARLYUPDATE/Resources/EAP-Brief-avian-flu.pdf.

16. R. Carlson, *The pace and proliferation of biological technologies,* Biosecur Bioterror 1, no. 3 (2003): 203–214; National Research Council, *Biotechnology research in a age of terrorism* (Washington, D.C.: National Academies Press, 2004).

17. See WHO, "Summary of probable SARS cases with onset of illness from 1 November 2002 to 31 July 2003," www.who.int/csr/sars/country/table2004_04_21/en/index.html.

18. R. S. Baric, *SARS-CoV: Lessons for global health,* Virus Res 133, no. 1 (2008): 1–3.

19. M. A. Marra, *The Genome sequence of the SARS-associated coronavirus,* Science 300, no. 5624 (2003): 1399–1404.

20. Ralph Baric, personal communication.

21. B. Yount, K. M. Curtis, and R. S. Baric, *Strategy for systematic assembly of large RNA and DNA genomes: Transmissible gastroenteritis virus model,* J Virol 74, no. 22 (2000): 10600–10611.

22. Z. Y. Yang et al., *A DNA vaccine induces SARS coronavirus neutralization and protective immunity in mice,* Nature 428, no. 6982 (2004): 561–564.

23. C. Fraser et al., *Factors that make an infectious disease outbreak controllable,* Proc Natl Acad Sci U.S.A. 101, no. 16 (2004): 6146–6151.

24. Ibid.

25. Robert Carlson, "Nature is full of surprises, and we are totally unprepared," www.synthesis.cc/2006/03/nature-is-full-of-surprises-and-we-are-totally-unprepared.html (first published on the Web on 5 March 2006).

26. H. Chen et al., *Establishment of multiple sublineages of H5N1 influenza virus in Asia: Implications for pandemic control,* Proc Natl Acad Sci U.S.A. 2006.

27. Ibid.

28. Ibid.

29. WHO, "Antigenic and genetic characteristics of H5N1 viruses and candidate H5N1 vaccine viruses developed for potential use as pre-pandemic vaccines," www.who.int/csr/disease/avian_influenza/guidelines/h5n1virus2006_08_18/en/index.html (accessed 18 August 2006).

30. E. Ghedin et al., *Large-scale sequencing of human influenza reveals the dynamic nature of viral genome evolution,* Nature 437, no. 7062 (2005): 1162–1166.

31. K. L. Russell et al., *Effectiveness of the 2003–2004 influenza vaccine among U.S. military basic trainees: A year of suboptimal match between vaccine and circulating strain,* Vaccine 23, no. 16 (2005): 1981.

32. On pediatric deaths, see CDC, "Update: Influenza activity—United States and worldwide, 2003–04 season, and composition of the 2004–05 influenza vaccine," *Morbidity and Mortality Weekly Report*, 2 July 2004, 547–552, www .cdc.gov/mmwr/preview/mmwrhtml/mm5325a1.htm; on jump in overall mortality, see CDC, "CDC 2003–04 U.S. influenza season summary," www.cdc .gov/flu/weekly/weeklyarchives2003–2004/03-04summary.htm.
33. S. Salzberg, *The contents of the syringe,* Nature 454, no. 7201 (2008): 160.
34. National Intelligence Council, "National Intelligence Estimate: The global infectious disease threat and its implications for the United States" (2000).
35. "SARS: Timeline of an outbreak," http://my.webmd.com/content/article/ 63/72068.htm; U.S. General Accounting Office, "West Nile virus outbreak: Lessons for public health preparedness," www.gao.gov/new.items/he00180 .pdf.
36. A. Mandavilli, *SARS epidemic unmasks age-old quarantine conundrum,* Nat Med 9, no. 5 (2003): 487.
37. R. G. Webster and E. A. Govorkova, *H5N1 influenza-continuing evolution and spread,* N Engl J Med 355, no. 21 (2006): 2174–2177; A. F. Oner et al., *Avian influenza A (H5N1) infection in eastern Turkey in 2006,* N Engl J Med 355, no. 21 (2006): 2179–2185; I. N. Kandun et al., *Three Indonesian clusters of H5N1 virus infection in 2005,* N Engl J Med 355, no. 21(2006): 2186–2194.
38. N. M. Ferguson et al., *Strategies for mitigating an influenza pandemic,* vol. 442 Nature (2006): 448–452.
39. C. Sheridan, *Production technologies change flu vaccine landscape,* Nat Biotechnol 25, no. 7 (2007): 701.
40. G. M. Forde, *Rapid-response vaccines—does DNA offer a solution?* Nat Biotechnol 23, no. 9 (2005): 1059–1062.
41. W. Gao et al., *Protection of mice and poultry from lethal H5N1 avian influenza virus through adenovirus-based immunization,* J Virol 80, no. 4 (2006): 1959–1964.
42. For additional details, see J. Newcomb, R. Carlson, and S. Aldrich, *Genome Synthesis and Design Futures: Implications for the U.S. Economy* (Cambridge, Mass.: Bio Economic Research Associates, 2007).
43. Ibid.
44. Carlson, *Pace and proliferation of biological technologies.*
45. E. D. Sevier and A. S. Dahms, *The role of foreign worker scientists in the US biotechnology industry,* Nat Biotechnol 20, no. 9 (2002): 955–956.
46. U.S. Department of Justice, usdoj.gov/dea/pubs/state_factsheets.html.
47. U.S. Drug Enforcement Administration, "Methamphetamine," www.usdoj.gov/ dea/concern/meth.html.
48. U.S. Drug Enforcement Administration, "Stats and facts," www.usdoj.gov/dea/ statistics.html.
49. U.S. Drug Enforcement Administration, "Maps of methamphetamine lab incidents," www.dea.gov/concern/map_lab_seizures.html.
50. See U.S. Drug Enforcement Administration, "National drug threat assessment 2007," www.usdoj.gov/dea/concern/18862/index.htm.

51. On the increase in meth use, see the U.S. Drug Enforcement Administration's statistics page at www.dea.gov/statistics.html; on the amount of meth seized at the border, see U.S. Drug Enforcement Administration, "Methamphetamine."

52. National Methamphetamine Threat Assessment 2007," *National Drug Intelligence Center,* November 2006, usdog.gov/ndic/pubs21/21821/overview.html.

53. *BBC News,* "Drug submarine found in Colombia," http://news.bbc.co.uk/2/hi/americas/915059.stm.

54. "Waving, not drowning," *Economist,* 3 May 2008.

55. See U.S. Department of Justice, "Methamphetamine strategic findings," www.usdoj.gov/dea/concern/18862/meth.htm#Strategic.

56. G. S. Pearson, *How to make microbes safer,* Nature 394, no. 6690 (1998): 217–218.

57. *The Kay Report to Congress on the activities of the Iraq Survey Group: Former bioweapons inspectors comment,* Biosecurity and Bioterrorism 1, no. 4 (2003): 239–246.

58. J. Huang et al., *Plant biotechnology in China,* Science 295, no. 5555 (2002): 674–678.

59. R. Stone, *Plant science: China plans $3.5 billion GM crops initiative,* Science 321, no. 5894 (2008): 1279.

60. H. Jia, *Chinese biotech hamstrung by production issues,* Nat Biotech 25, no. 2 (2007): 147.

61. H. Breithaupt, *China's leap forward in biotechnology,* EMBO Rep 4, no. 2 (2003): 111–113.

62. S. Pearson, H. Jia, and K. Kandachi, *China approves first gene therapy,* Nat Biotechnol 22, no. 1 (2004): 3–4.

63. G. Epstein, *Global Evolution of Dual-Use Biotechnology* (Washington, D.C.: Center for Strategic and International Studies, 2005); Charles Cantor, "Global evolution of dual-use biotechnology: 2020" (presentation at the National Intelligence Council Conference, Center for Strategic and International Studies, Washington, D.C., 18 March 2004).

64. A. R Taylhardat and A. Falaschi, *Funding assured for India's international biotechnology centre,* Nature 409, no. 6818 (2001): 281; "India commits to boosting biotechnology research," Nature 450, no. 599 (2007); K. S. Jayaraman, *India promotes GMOs in Asia,* Nat Biotechnol 20, no. 7 (2002): 641–642.

65. *Singapore attracts foreign talent,* Nature 394 (1998): 604.

66. D. Swinbanks and D. Cyranoski, *Taiwan backs experience in quest for biotech success,* Nature 407, no. 6802 (2000): 417–426; D. Cyranoski, *Taiwan: Biotech vision,* Nature 421, no. 6923 (2003): 672–673.

67. H. Bügl et al., *DNA synthesis and biological security,* Nat Biotech 25, no. 6 (2007): 627.

68. C. Dreifus, "A conversation with: Robert C. Richardson; The chilling of American science," *New York Times,* 6 July 2004.

10. The Sources of Innovation

1. W. B. Arthur, *The structure of invention*, Research Policy 36, no. 2 (2007): 274.

2. W. Baumol, *Small Firms: Why Market-Driven Innovation Can't Get Along without Them* (Washington, D.C.: U.S. Small Business Administration, 2005), 183.

3. This particular list is from Baumol, *Small Firms,* from the original in U.S. Small Business Administration, Office of Advocacy, *The State of Small Business: A Report to the President* (Washington, D.C.: Government Printing Office, 1994).

4. National Science Foundation, Division of Science Resources Statistics, "Research and development in industry: 2003," NSF 07–314 (2006), www.nsf.gov/statistics/nsf07314/start.cfm p30.

5. National Research Council, *Funding a Revolution: Government Support for Computing Research* (Washington, D.C.: National Academies Press, 1999).

6. "Less glamour, more profit," *Economist,* 22 April 2004.

7. Baumol, *Small Firms,* 187.

8. Ibid.

9. Peregrine Analytics, *Innovation and Small Business Performance: Examining the Relationship between Technological Innovation and the Within-Industry Distributions of Fast Growth Firms* (Washington, D.C.: U.S. Small Business Adminstration, 2006).

10. Baumol, *Small Firms,* 199.

11. Ibid, 184.

12. J. Randerson, "Lax laws, virus DNA and potentital for terror," *Guardian,* 14 June 2006; J. Randerson, "Revealed: The lax laws that could allow assembly of deadly virus DNA," *Guardian,* 14 June 2006.

13. Randerson, "Revealed."

14. Ibid.

15. "Newspaper investigation highlights bioterror fears," In the Field: The *Nature* Reporters' Blog from Conferences and Events, *Nature.com,* 14 June 2006, http://blogs.nature.com/news/blog/2006/06/newspaper_investigation_highli.html.

16. See "Revised NSABB charter," signed 16 March 2006, http://www.biosecurityboard.gov/revised%20NSABB%20charter%20signed%2003 1606.pdf.

17. See National Select Agents Registry, "Frequently asked questions," http://www.selectagents.gov/NSARFaq.htm.

18. See Center for Infectious Disease Research and Policy, "Smallpox summary" page, http://www.cidrap.umn.edu/cidrap/content/bt/smallpox/biofacts/smllpx-summary.html.

19. M. Enserink, *BIODEFENSE: Unnoticed amendment bans synthesis of smallpox virus,* Science 307, no. 5715 (2005): 1540a–1541a.

20. Y. Bhattacharjee, *Smallpox law needs fix,* Science NOW, 25 October 2006.

21. Enserink, *BIODEFENSE,* 1541.

22. National Science Advisory Board for Biosecurity, *Addressing Biosecurity Concerns Related to the Synthesis of Select Agents* (2006), 12, oba.od.nih.gov/biosecurity/biosecurity_documents.html.

23. Committee on Advances in Technology and the Prevention of Their Application to Next Generation Biowarefare Threats, National Research Council, *Globalization, Biosecurity, and the Future of the Life Sciences* (Washington, D.C.: National Academies Press, 2006), 218.
24. Full disclosure: I briefed this committee during deliberations and served as an academic reviewer prior to publication.
25. National Science Advisory Board for Biosecurity, *Addressing Biosecurity Concerns,* 12.
26. Full disclosure: I briefed this committee during deliberations and served as an academic reviewer prior to publication. M. Garfinkel et al., *Synthetic Genomics: Options for Governance* (Rockville, Md.: J. Craig Venter Institute, in collaboration with CSIS, MIT, 2007), 6; quotations, i, ii.
27. Ibid.
28. Ibid., 27.
29. *BBC News,* "Bugging the boardroom," 5 September 2006, http://news.bbc .co.uk/2/hi/technology/5313772.stm.
30. J. Newcomb, R. Carlson, and S. Aldrich, *Genome Synthesis and Design Futures: Implications for the U.S. Economy* (Cambridge, Mass.: Bio Economic Research Associates, 2007).
31. Garfinkel et al., *Synthetic Genomics.*
32. Note that the probable emergence of a market for such an instrument is a hypothesis, and time will provide the experimental test.
33. R. Carlson, *The pace and proliferation of biological technologies,* Biosecur Bioterror 1, no. 3 (2003): 203–214.
34. K. Kelly, "The futilitiy of prohibitions," 2006, available at www.kk.org/ thetechnium/archives/2006/02/the_futility_of.php.
35. Ibid.

11. Laying the Foundations for a Bioeconomy

1. See the USDA report, "Productivity growth in U.S. agriculture," www.ers .usda.gov/publications/eb9.
2. J. Hodgson, *Private biotech 2004—the numbers,* Nat Biotech 24, no. 6 (2006): 635.
3. See the CIA *World Factbook,* www.cia.gov/library/publications/the-world -factbook/index.html.
4. M. Herper and P. Kang, "The world's ten best-selling drugs," *Forbes,* 22 March 2006; S. Aggarwal, *What's fueling the biotech engine?* Nat Biotech 25, no. 10 (2007): 1097; Herper and Kang, "The world's ten best-selling drugs."
5. Herper and Kang, "The world's ten best-selling drugs"; Aggarwal, *What's fueling the biotech engine?*
6. A. Berenson, "Weak sales prompt Pfizer to cancel diabetes drug," *New York Times,* 19 October 2007.
7. Ibid.; Aggarwal, *What's fueling the biotech engine?*; "TOP 20 Biologics 2008," R&D Pipeline News, La Merie Business Intelligence, www.bioportfolio .com.

8. Ibid.
9. V. Ozdemir et al., *Shifting emphasis from pharmacogenomics to theragnostics,* Nat Biotech 24, no. 8 (2006): 942.
10. Ibid.
11. "Beyond the blockbuster."
12. Ibid.
13. The latest data is from 2004, and the U.S. economy has been growing at about 4 percent per year. See "The statistical abstract," www.census.gov/compendia/statab.
14. D. Pimentel, R. Zuniga, and D. Morrison, *Update on the environmental and economic costs associated with alien-invasive species in the United States,* Ecological Economics 52, no. 3 (2005): 273.
15. S. Lawrence, *Agbiotech booms in emerging nations* Nat Biotech 25, no. 3 (2007): 271.
16. For percentages of GM crops, see "Adoption of Genetically Engineered Crops in the U.S.," Economic Research Service, USDA, www.ers.usda.gov/Data/BiotechCrops/; for total crop revenues, see the "Data and Statistics" page, National Agricultural Statistics Service, USDA, www.nass.usda.gov/Data_and_Statistics/Quick_Stats/index.asp.
17. See Hepeng Jia, "Big funding for GM research," *Chemistry World,* 26 March 2008, www.rsc.org/chemistryworld/News/2008/March/26030801.asp.
18. James, Clive, "Global Status of Commercialized Biotech/GM Crops: 2008," International Service for the Acquisition of Agri-Biotech Applications, ISAAA Brief No. 39. ISAAA: Ithaca, NY.
19. "The next green revolution," *Economist,* 21 February 2008.
20. See, for example, Biotechnology Industry Organization, "Agricultural production applications," http://bio.org/speeches/pubs/er/agriculture.asp.
21. P. Mitchell, *GM giants pair up to do battle,* Nat Biotech 25, no. 7 (2007): 695.
22. P. Mitchell, *Europe's anti-GM stance to presage animal feed shortage?* Nat Biotech 25, no. 10 (2007): 1065.
23. J. L. Fox, *US courts thwart GM alfalfa and turf grass,* Nat Biotech 25, no. 4 (2007): 367.
24. Pimentel, Zuniga, and Morrison, *Update on the environmental and economic costs associated with alien-invasive species in the United States.*
25. "Industrial biotechnology—turning potential into profits," *EuropaBio Bio-Economy Newsletter,* September 2006.
26. S. Herrera, *Industrial biotechnology—a chance at redemption,* Nat Biotech 22, no. 6 (2004): 671; "Field of dreams," *Economist,* 7 April 2004.
27. "Industrial Biotechnology—Turning Potential into Profits," EuropaBio Bio-Economy Newsletter, September, 2006; Riese, Jense, "White Biotechnology Press Briefing," McKinsey & Company, February 2009.
28. S. Aldrich, J. Newcomb, and R. Carlson, *The Big Squeeze: New Fundamentals for Food and Fuel Markets* (Cambridge, Mass.: Bio Economic Research Associates, 2008).
29. P. Westcott, "U.S. ethanol expansion driving changes throughout the agricultural sector," *Amber Waves,* September 2007.

30. J. Newcomb, "Chomp! Chomp! Fueling the new agribusiness," *CLSA Blue Books,* April 2007.
31. Ibid.
32. K. Fuglie, "Productivity drives growth in U.S. agriculture," *Amber Waves,* September 2007.
33. W. Bi, "Increasing domestic demand could cut China's corn exports," Bloomberg via *Livemint.com/Wall Street Journal,* 24 March 2007, www .livemint.com/2007/03/24022323/Increasing-domestic-demand-cou.html.
34. See Green Car Congress, "China halts expansion of corn ethanol industry: Focus on biomass feedstocks," www.greencarcongress.com/2006/12/china _halts_exp.html.
35. A. Barrionuevo, "Boom in ethanol reshapes economy of heartland," *New York Times,* 25 June 2006.
36. "Cheap no more," *Economist,* 6 December 2007.
37. C. Krauss, "Ethanol's boom stalling as glut depresses price," *New York Times,* 30 September 2007.
38. P. Barta, "Promising crop: Jatropha plant gains steam in global race for biofuels; Hardy shrub is tapped for energy-rich seeds; Indian farmers' big bet," *Wall Street Journal,* 24 August 2007.
39. "China to subsidize jatropha planting for biodiesel," Reuters, 5 June 2007.
40 A. Hind, "Could jatropha be a biofuel panacea?" *BBC News,* 8 July 2007.
41. For a simple description of the structure of the plant cell wall, see "Fuel ethanol production" from the Genomes to Life (GTL) program at the U.S. Department of Energy, http://genomicsgtl.energy.gov/biofuels/ethanolproduction.shtml.
42. See "BP, ABF and DuPont unveil $400 million investment in UK biofuels," www.bp.com/genericarticle.do?categoryId=2012968&contentId=7034350.
43. R. Chase, "DuPont, BP join to make butanol: They say it outperforms ethanol as a fuel additive," *USA Today,* 20 June 2006.
44 See, U.S. Department of Energy, "DOE selects six cellulosic ethanol plants for up to $385 million in federal funding," 28 February 2007, www.doe.gov/news/4827.htm.
45. P. Grace et al., *The potential impact of agricultural management and climate change on soil organic carbon of the north central region of the United States,* Ecosystems 9, no. 5 (2006).
46. S. M. Ogle et al., *Uncertainty in estimating land use and management impacts on soil organic carbon storage for US agricultural lands between 1982 and 1997,* Global Change Biology 9, no. 11 (2003): 1521–1542.
47. P. G. Johnson et al., *Pollen-mediated gene flow from Kentucky bluegrass under cultivated field conditions,* Crop Science 46, no. 5 (2006): 1990; L. S. Watrud et al., *From the cover: Evidence for landscape-level, pollen-mediated gene flow from genetically modified creeping bentgrass with CP4 EPSPS as a marker,* PNAS 101, no. 40 (2004): 14533; J. R. Reichman et al., *Establishment of transgenic herbicide-resistant creeping bentgrass (Agrostis stolonifera L.) in nonagronomic habitats,* Molecular Ecology 15, no. 13 (2006): 4243.
48 S. I. Warwick et al., *Do escaped transgenes persist in nature? The case of an herbicide resistance transgene in a weedy* Brassica rapa *population,* Molecular Ecology 17, no. 5 (2007): 1387–1395.

49. I. A. Zelaya, M. D. K. Owen, and M. J. VanGessel, *Transfer of glyphosate resistance: Evidence of hybridization in* Conyza (Asteraceae), American Journal of Botany 94, no. 4 (2007): 660.

50. D. Tilman et al., *Diversity and productivity in a long-term grassland experiment*, Science 294, no. 5543 (2001): 843.

51. D. Tilman, J. Hill, and C. Lehman, *Carbon-negative biofuels from low-input high-diversity grassland biomass*, Science 314, no. 5805 (2006): 1598.

52. M. R. Schmer et al., *Net energy of cellulosic ethanol from switchgrass*, PNAS 105, no. 2 (2008): 464–469.

53. E. A. Heaton, "Miscanthus bioenergy: Achieving the 2015 yield goal of switchgrass" (paper presented at the Biomass Symposium, University of Illinois at Urbana-Champaign, 2007).

54. "Hybrid grass may prove to be valuable fuel source," *ScienceDaily*, 30 September 2005.

55. C. Wyman, "More on biofuels discussion," *Chemical and Engineering News*, 24 March 2008; U.S. Energy Information, "Petroleum basic statistics," www.eia.doe.gov/basics/quickoil.html; R. Lubowski et al., "Major uses of land in the United States, 2002," *Economic Information Bulletin*, no. EIB-14, May 2006, USDA Economic Research Service.

56. J. Fargione et al., *Land clearing and the biofuel carbon debt*, Science 319, no. 5867 (2008): 1235–1238.

57. T. Searchinger et al., *Use of U.S. croplands for biofuels increases greenhouse gases through emissions from land-use change*, Science 319, no. 5867 (2008): 1238–1240.

58. See "New studies portray unbalanced perspective on biofuels: DOE committed to environmentally sound biofuels development," 23 May 2008, and "DOE response to *Science* magazine article [DOE actively engaged in investigating the role of biofuels in greenhouse gas emissions from indirect land use change]," www1.eere.energy.gov/biomass/news_detail.html?news_id=11794.

59. "Cheap no more."

60. E. Singer, "Greener jet fuel," *Technology Review*, 11 June 2007.

61. D. Phillips, "Air Force hopes to cut oil's role in fuel," *New York Times*, 18 June 2007.

62. Z. Serber, "Artemisinin and biofuel production" (paper presented at the Fourth Annual International Conference on Synthetic Biology, Hong Kong University of Science and Technology, 10–12 October 2008).

63. E. Seba, "Shell, Saudi commit to massive U.S. refinery project," Reuters, 21 September 2007.

64. See "BP, ABF and DuPont unveil $400 million investment in UK biofuels," www.bp.com/genericarticle.do?categoryId=2012968&contentId=7034350.

65. S. Atsumi, T. Hanai, and J. C. Liao, *Non-fermentative pathways for synthesis of branched-chain higher alcohols as biofuels*, Nature 451, no. 7174 (2008): 86–89.

66. R. Carlson, *Open-source biology and its impact on industry*, IEEE Spectrum (May 2001).

67. "Corn used as raw material for plastic bottle and fabric," *EngineerLive*, 1 October 2006, www.engineerlive.com/european-chemical-engineer/safety

-in-the-plant/13234/corn-used-as-raw-material-for-plastic-bottles-and-fabrics .thtml.

68. Y. H. P. Zhang, *High-yield hydrogen production from starch and water by a synthetic enzymatic pathway,* PLoS ONE, no. 5 (2007): e456.
69. For a more in-depth review, see J. Newcomb, R. Carlson, and S. Aldrich, *Genome Synthesis and Design Futures: Implications for the U.S. Economy* (Cambridge, Mass.: Bio Economic Research Associates, 2007).
70. W. B. Arthur, *The structure of invention,* Research Policy 36, no. 2 (2007): 274.
71. S. Berkun, *The Myths of Innovation* (Sebastopol, Calif.: O'Reilly, 2007).
72. Newcomb, Carlson, and Aldrich, *Genome Synthesis and Design Futures.*

12. Of Straitjackets and Springboards for Innovation

1. My focus here on U.S. intellectual property law is by no means meant to reduce the importance of law or technology development in other countries or of international agreements. In the short term, U.S. law is important to understand because the lion's share of revenues from genetically modified systems are presently in the United States, as described in Chapter 11, and will continue to be so for some years to come. However, U.S. laws and legal precedents are also important to focus on precisely because the rest of the world will, in the future, face no economic or technical barriers to directly competing with the United States in developing new biological technologies and in deriving subsequent economic benefit. Whether intellectual property regimes in the United States foster or hinder biological innovation will in large part determine the future ability of the country to compete on the world stage.
2. The U.S. Constitution is available online via the U.S. Government Printing Office, www.gpoaccess.gov/constitution/index.html.
3. See U.S. Patent and Trademark Office, "General information concerning patents," www.uspto.gov/go/pac/doc/general.
4. See "The United States Patent Office—sketch of its history," *New York Times,* 25 July 1859, http://query.nytimes.com/gst/abstract.html?res=9D01EFDA1F 31EE34BC4D51DFB1668382649FDE.
5. L. Lessig, *The Future of Ideas* (New York: Random House, 2001), 212, 259.
6. Ibid., 236.
7. M. W. McFarland, ed. *The Papers of Wilbur and Orville Wright, Including the Chanute-Wright Letters and Other Papers of Octave Chanute* (New York: McGraw-Hill, 1953), 17.
8. Ibid., 19.
9. See U.S. Centennial of Flight Commission, "Glenn Curtiss and the Wright patent battles," www.centennialofflight.gov/essay/Wright_Bros/Patent_Battles/ WR12.htm.
10. McFarland, *Papers of Wilbur and Orville Wright,* 982.
11. J. Clarke et al., *Patent Pools: A Solution to the Problem of Access in Biotechnology Patents?* (Washington D.C.: United States Patent and Trademark Office, 2000).
12. Ibid., 4.

13. R. Merges, *Institutions for Intellectual Property Transactions: The Case of Patent Pools* (1999). Preprint available at: www.berkeley.edu/institutes/belt/pubs/merges/pools.pdf.

14. Ibid., 21.

15. M. A. Heller and R. S. Eisenberg, *Can patents deter innovation? The anticommons in biomedical research,* Science 280, no. 5364 (1998): 698.

16. Merges, *Institutions for Intellectual Property Transactions.*

17. J. Newcomb, R. Carlson, and S. Aldrich, *Genome Synthesis and Design Futures: Implications for the U.S. Economy* (Cambridge, Mass.: Bio Economic Research Associates, 2007).

18. See U.S. Patent and Trademark Office, "USPTO issues white paper on patent pooling," 19 January 2001, www.uspto.gov/web/offices/com/speeches/01-06.htm.

19. Clarke et al., *Patent Pools,* 6.

20. Merges, *Institutions for Intellectual Property Transactions.*

21. A. Rai and J. Boyle, *Synthetic biology: Caught between property rights, the public domain, and the commons,* PLoS Biology 5, no. 3 (2007): e58.

22. Ibid.

23. Newcomb, Carlson, and Aldrich, *Genome Synthesis and Design Futures,* 49.

24. A. Yancey and C. N. Stewart, *Are university researchers at risk for patent infringement?* Nat Biotech 25, no. 11 (2007): 1225.

25. R. Carlson, *Open-source biology and its impact on industry,* IEEE Spectrum (May 2001); R. Carlson, *The pace and proliferation of biological technologies,* Biosecur Bioterror 1, no. 3 (2003): 203–214; R. Carlson, "Splice it yourself," *Wired,* May 2005.

26. J. Paradise, L. Andrews, and T. Holbrook, *Intellectual property: Patents on human genes: An analysis of scope and claims,* Science 307, no. 5715 (2005): 1566.

27. McFarland, *Papers of Wilbur and Orville Wright,* 989.

28. A. Davidson, "Big firms eye 'open innovation' for ideas," *Weekend Edition Sunday,* National Public Radio, 27 May 2007, www.npr.org/templates/story/story.php?storyId=10480377.

29. R. Buderi, *What new economy?* Technology Review (January 2001): 45–50.

30. Y. Benkler, *The Wealth of Networks.* (New Haven, Conn.: Yale University Press, 2006), 40. The Wealth of Networks is available online at www.benkler.org/wealth_of_networks/index.php/Main_Page.

31. See the IBM 2006 annual report, http://www.ibm.com/annualreport/2006/cfs_earnings.shtml.

32. V. Vaitheeswaran, "The love-in," *Economist,* 11 October 2007.

33. Ibid.

34. H. Chesbrough, *Open Innovation: The New Imperative for Creating and Profiting from Technology* (Cambridge, Mass.: Harvard Business School Press, 2003).

35. Ibid., xx.

36. Ibid., xxii.

37. Ibid., xxxiv.

38. Vaitheeswaran, "The love-in."

39. Chesbrough, *Open Innovation.*
40. Vaitheeswaran, "The love-in."
41. Y. Benkler, *Coase's penguin, or, Linux and the nature of the firm,* Yale Law Journal 112, no. 3 (2002): 371.
42. Ibid., 369.
43. S. Weber, *The Success of Open Source* (Cambridge, Mass.: Harvard University Press, 2004), 84.
44. Ibid., 9.
45. Ibid., 1.
46. J. Hope, *Biobazaar: The Open Source Revolution and Biotechnology* (Cambridge, Mass.: Harvard University Press, 2008).
47. See PageRank explained at www.google.com/technology.
48. S. Stokely, "Hardware vendors will follow the money to open source," *ITNews,* 31 January 2008, www.itnews.com.au/News/69229,hardware-vendors -will-follow-the-money-to-open-source.aspx.
49. See the Netcraft monthly Web Server Survey, http://news.netcraft.com/ archives/web_server_survey.html.
50. Vaitheeswaran, "The love-in."
51. R. A. Ghosh, *Economic Impact of Open Source Software on INNOVATION and the Competitiveness of the information and communication technologies (ICT) sector in the EU* (Brussels: European Commission, Directorate General for Enterprise and Industry, 2007).
52. See GNU Operating System, "The free software definition," www.gnu.org/ philosophy/free-sw.html.
53. See Wikipedia, "GNU General Public License" or "GPL," http://en.wikipedia .org/wiki/Gpl.
54. J. Buchanan, "Linus Tovalds talks future of Linux," *APC Magazine,* 22 August 2007.
55. B. Perens, "The emerging economic paradigm of open source," last edited, 16 Feb 2005; accessed, 12 Jan 2008, http://perens.com/Articles/Economic.html.
56. W. Baumol, *Small Firms,* from the original in U.S. Small Business Administration, Office of Advocacy, *The State of Small Business: A Report to the President* (Washington, D.C.: Government Printing Office, 1994), 195.
57. Perens, "The emerging economic paradigm of open source."
58. Carlson, *Open-source biology and its impact on industry.*

13. Open-Source Biology, or Open Biology?

1. See, "DARPA open-source biology letter," www.synthesis.cc/DARPA_OSB _Letter.html.
2. The Biobrick Foundation, "Our goals," as of 9 January 2008, www .biobricks.org.
3. Drew Endy, personal communication.
4. A. Rai and J. Boyle, *Synthetic biology: Caught between property rights, the public domain, and the commons,* PLoS Biology 5, no. 3 (2007): e58.
5. S. Weber, *The Success of Open Source* (Cambridge, Mass.: Harvard University Press, 2004), 84.

6. "Open source biotech," *Red Herring,* April 2006.
7. See "BiOS license and tech support agreement version 1.5," www.bios.ret/daisy/bios/mta.
8. "Open source biotech," 32.
9. See, for example, "BiOS mutual non-assertion agreement, v2.0," www.bios.ret/daisy/bios/mta.
10. See "BiOS PMET license agreement," version 1.5, also called "The CAMBRIA biological open source (BiOS) license for plant enabling technologies," version 1.5, www.bios.ret/daisy/bios/mta.
11. See "What is the cost of a BiOS agreement?" at www.bios.net/daisy/bios/licenses/398/2535.html.
12. L. Lessig, *The Future of Ideas* (New York: Random House, 2001), 236.
13. E. Raymond, *The Cathedral and the Bazaar: Musings on Linux and Open Source by an Accidental Revolutionary* (Cambridge, Mass.: O'Reilly Media, 2001). The text is online at http://catb.org/~esr/writings/cathedral-bazaar/ cathedral-bazaar.
14. J. Hope, *Biobazaar: The Open Source Revolution and Biotechnology* (Cambridge Mass.: Harvard University Press, 2008).
15. Ibid., 104.
16. J. Hope, "Open source biotechnology?" http://rsss.anu.edu.au/~janeth/OSBiotech.html.
17. Hope, *Biobazaar,* 139.
18. Hope, "Open source biotechnology?"
19. Ibid.
20. See S. Weber, *The Success of Open Source* (Cambridge, Mass.: Harvard University Press, 2008), 99.
21. Hope, *Biobazaar,* 139.
22. See J. Hope, "Open source biotechnology: A new way to manage scientific intellectual property," GeneWatch Magazine 18(1) 2005.
23. See Yahoo Finance, "Google Inc. (GOOG)," http://finance.yahoo.com/q/is?s=GOOG&annual.
24. S. Stokely, "Hardware vendors will follow the money to open source," *ITNews,* 31 January 2008, http://www.itnews.com.au/News/69229,hardware-vendors-will-follow-the-money-to-open-source.aspx (accessed 2 February 2008).
25. Hope, *Biobazaar,* 71.
26. Estimates from the FBI, Interpol, World Customs Organization, and the International Chamber of Commerce. See http://www.stopfakes.gov/sf_why.asp.
27. Ibid.
28. Hope, *Biobazaar,* 150.
29. D. Byrne, "David Byrne's survival strategies for emerging artists—and megastars, *Wired,* 18 December 2007.
30. Ibid.
31. David Byrne, in conversation with Brian Eno, "How the f—k can we get out of this?" www.wired.com/entertainment/music/magazine/16-01/ff_byrne?currentPage=2.
32. Byrne, "David Byrne's survival strategies."
33. See "Digital Millennium Copyright Act," http://en.wikipedia.org/wikiDigital_Millennium_Copyright_Act (accessed 28 January 2008).

14. What Makes a Revolution?

1. D. Bennett, "Environmental defense," *Boston Globe,* 27 May 2007.
2. See the "Oil and the military" page from the Lugar Energy Initiative, http://lugar.senate.gov/energy/security/military.cfm.
3. On the DoD's rank in terms of fuel use, see Sohbet Karbuz, "US military oil pains," Energy Bulletin, 17 February 2007, http://www.energybulletin.net/26194.html.
4. W. Murray and M. Knox, "Thinking about Revolutions in Warfare," in M. Knox and W. Murray, eds., *The Dynamics of Military Revolutions: 1300–2050* (Cambridge: Cambridge University Press, 2001).
5. M. Grimsley, "Surviving Military Revolutions: The U.S. Civil War," in M. Knox and W. Murray, eds., *The Dynamics of Military Revolution: 1300–2050* (Cambridge: Cambridge University Press, 2001).
6. Murray and Knox, "Thinking about Revolutions in Warfare."
7. Ibid.
8. W. Murray and M. Knox, "The Future Is behind Us," in M. Knox and W. Murray, eds., *The Dynamics of Military Revolutions: 1300–2050* (Cambridge: Cambridge University Press, 2001).
9. J. Cascio, "The lost hegemon (pt 2): The end of conventional war," Open the Future [blog], 7 May 2007, www.openthefuture.com/2007/05/the_lost_hegemon_pt_2_the_end.html (accessed 2 January 2008).
10. J. W. Anderson, S. Fainaru, and J. Finer, "Bigger, stronger homemade bombs now to blame for half of U.S. deaths," *Washington Post,* 16 October 2005; C. Wilson, *Improvised Explosive Devices in Iraq: Effects and Countermeasures* (Washington D.C.: Congressional Research Service, Library of Congress, 2005).
11. M. Enserink, *The anthrax case: From spores to a suspect,* ScienceNOW, 12 August 2008.
12. On IEDs, see Anderson, "Bigger, stronger homemade bombs."
13. See, for example, DIY Drones, http://diydrones.com.
14. Chris Anderson, the editor of *Wired* magazine, posted an entry on the DIY Drones community blog titled "Can open source be giving comfort to the enemy?" at www.longtail.com/the_long_tail/2007/08/can-open-source.html. Anderson notes that he is "honestly conflicted" about misuse of the technology but is inclined to help anyone who asks, despite the community being "just a pen stroke away from being regulated out of existence, and in this climate it's politically unwise to discount the Homeland Security card ([his] own feelings about that notwithstanding)." Comments from other participants in the community are wholly supportive of sharing.
15. For list of tools, see Fab Central, http://fab.cba.mit.edu/content/tools; for quotation, see the Fab Lab wiki, http://fab.cba.mit.edu/about/about2.php.
16. See the Fab@Home main page, http://fabathome.org/wiki/index.php?title=Main_Page.
17. A. Anderson, "A whole new dimension," *Economist: The World in 2008,* 2007.
18. See U. Hedquist, "Open source 3D printer copies itself," *Computerworld,* 8 April 2008, http://computerworld.co.nz/news.nsf/tech/2F5C3C5D68A380ED

CC257423006E71CD, and RepRap, "What is RepRap?" www.reprap.org/bin/view/Main/WebHome.

19. V. Vaitheeswaran, "A dark art no more," *Economist,* 11 October 2007.

20. Ibid.; "Revving up," *Economist,* 11 October 2007.

21. Peregrine Analytics, *Innovation and Small Business Performance: Examining the Relationship between Technological Innovation and the Within Industry Distributions of Fast Growth Firms* (Washington, D.C.: U.S. Small Business Adminstration, 2006), 7.

22. Ibid.

23. On amounts spent on R&D, see PhRMA, www.phrma.org/about_phrma; for quotation, see "Don't laugh at gilded butterflies," *Economist,* 22 April 2004.

24. G. Pisano, "Can science be a business? Lessons from Biotech," *Harvard Business Review,* 1 October 2006.

25. "Beyond the blockbuster," *Economist,* 28 July 2007.

26. Pisano, *Can science be a business?,* 115, 119.

27. D. Edgerton, *The Shock of the Old: Technology and Global History Since 1900* (New York: Oxford University Press, 2007), xv.

28. Ibid., 79.

29. Ibid., 80.

30. Ibid.

31. Ibid., 77.

32. Ibid., chap. 4.

33. See "Security in ten years," Schneier on Security, 3 December 2007, www.schneier.com/blog/archives/2007/12/security_in_ten.html.

34. R. Carlson, *Open-source biology and its impact on industry,* IEEE Spectrum (May 2001).

35. K. Amanuma et al., *Transgenic zebrafish for detecting mutations caused by compounds in aquatic environments,* Nat Biotechnol 18, no. 1 (2000): 62–65; H. E. David et al., *Construction and evaluation of a transgenic hsp16-GFP-lacZ* Caenorhabditis elegans *strain for environmental monitoring,* Environ Toxicol Chem 22, no. 1 (2003): 111–118; L. Nelson, "Plants to uncover landmines," *Nature News,* 29 January 2004.

36. See remarks of Senator Bill Frist at Harvard Medical School, 1 June 2005, www.synthesis.cc/2005/06/bill_frists_bio.html; see also R. Kurzweil and B. Joy, "Recipe for destruction," *New York Times,* 17 October 2005, www.nytimes.com/2005/10/17/opinion/17kurzweiljoy.html.

37. H. Pearson et al., *SARS: What have we learned?* Nature 424, no. 6945 (2003): 121–126.

Afterword

1. See the iGEM 2008, http:2008.igem. org/Main_Page.

2. Michelle M. Becker, et al., *Synthetic recombinant bat SARS-like coronavirus is infectious in cultured cells and in mice,* PNAS 105, no. 50 (December 16, 2008): 19944–19949.

3. See the press release http:ucsusa.org/food_and_agriculture/science_and_impacts/ science/failure-to-yield.html; and Doug Gurian-Sherman, *Failure to Yield: Evaluating the Performance of Genetically Engineered Crops* (Cambridge, Mass.: UCS Publications, 2009).

4. See "BIO Debunks Myths in Anti-Industry Report: Agricultural Biotechnology Helps Farmers Increase Crop Production," www.bio.org/news/pressreleases/ newsitem.asp?id=2009_0414_01.

Index